Magnets and Motors

TEACHER'S GUIDE

SCIENCE AND TECHNOLOGY FOR CHILDREN

NATIONAL SCIENCE RESOURCES CENTER
Smithsonian Institution—National Academy of Sciences
Arts and Industries Building, Room 1201
Washington, DC 20560

NSRC

The National Science Resources Center is operated by the Smithsonian Institution and the National Academy of Sciences to improve the teaching of science in the nation's schools. The NSRC collects and disseminates information about exemplary teaching resources, develops and disseminates curriculum materials, and sponsors outreach activities, specifically in the areas of leadership development and technical assistance, to help school districts develop and sustain hands-on science programs. The NSRC is located in the Arts and Industries Building of the Smithsonian Institution in Washington, D.C.

Copyright © 1991, National Academy of Sciences. All rights reserved.
Reprinted 1992.

Published by Carolina Biological Supply Company, 2700 York Road, Burlington, NC 27215.
Call toll free 800-334-5551. In North Carolina, call 800-632-1231.

No part of this book may be reproduced by any mechanical, photographic, or electronic process, or in the form of a phonographic recording, nor may it be stored in a retrieval system, transmitted, or otherwise copied for public or private use without permission in writing from the National Science Resources Center.

NSRC Staff

Douglas Lapp, Executive Director
Sally Shuler, Deputy Director
Gail Greenberg, Executive Administrative Assistant
Charmane Beverly, Program Assistant
Mary Garino, Office Aide
Karen Fusto, Administrative Officer
Diane Mann, Administrative Assistant
Kathleen Johnston, Publications Director
Lynn Miller, Writer/Editor
Marilyn Fenichel, Writer/Editor
Catherine Corder, Publications Technology Specialist
Max-Karl Winkler, Illustrator
Lois Sloan, Illustrator Consultant
Heidi Kupke, Publications Assistant
Joyce Weiskopf, STC Project Director
Wendy Binder, STC Research Associate
Debby Deal, STC Research Associate
David Hartney, STC Research Associate
Patricia McGlashan, STC Research Associate
Katherine Stiles, STC Research Associate
Laura Pierce, Program Assistant
Jan Tuomi, Outreach Director
Betty Olivolo, Program Officer
Catherine Harris, Program Assistant
Patricia McClure, Information Dissemenation Project Director
Ted Schultz, Program Officer
Ramon Lopez, Networking Consultant
Maria Mims, Program Assistant

Foreword

Why study science? Science, more than any other discipline, provides us with tools to learn about the world. Science is not a listing of facts; science is an invitation to observe the world, ask questions, and puzzle over problems and enjoy the process of solving them. From the time children begin to perceive their environment, they are involved in science. This kind of science is the core of the NSRC's Science and Technology for Children (STC) project; its goal is to provide teachers and children with an inquiry-based curriculum that builds on children's excitement as they discover new things and guides them in arriving at explanations that make sense to them.

This unit is one of 24 elementary science curriculum units being developed by the STC project for grades one through six. Each STC unit provides children with the opportunity to learn in-depth about topics in the physical, life, or earth sciences and technology through direct observation and experimentation. The units invite children to develop hypotheses, then to test their ideas, just as professional scientists do. Discovering what makes a light bulb turn on and the way plants grow are just as exciting—and important—events to young children as the discoveries that scientists make. Along the way, children develop patience, persistence, and confidence in their own ability to tackle and solve real problems.

As the teacher, your role in this process is crucial. You will be guiding hands-on learning—encouraging students to explore new ideas for themselves. This is a rewarding way of teaching—and exciting. You will be helping students learn to think for themselves as they expand their understanding of the world around them.

Acknowledgments

The National Science Resources Center would like to acknowledge the many individuals and organizations that have supported the Science and Technology for Children project. The major funding for the project is from the John D. and Catherine T. MacArthur Foundation, with additional funds from the Dow Chemical Company Foundation; E.I. du Pont de Nemours and Company; Amoco Foundation, Inc.; the U.S. Department of Defense; and the U.S. Department of Education.

The primary author of the *Magnets and Motors* unit was David Hartney. In developing the unit, he worked closely with a number of schools, including the Stuart-Hobson School of the Capitol Hill Cluster Schools in Washington, D.C., where he trial-taught the unit with Catherine Taylor.

The following individuals also contributed to the development and pilot testing of the unit:

Brent Atkins, Rockbridge Elementary School, Norcross, GA

L. J. Benton, Coordinator, Instructional Materials Processing Center, Fairfax County Public Schools, Fairfax, VA

Lila Bishop, Sidwell Friends School, Bethesda, MD

Bernard Finn, Curator, Division of Electricity and Modern Physics, National Museum of American History, Smithsonian Institution

Susan Glenn, Assistant Archivist, Smithsonian Archives, Smithsonian Institution

David Green, C.E. Jordan High School, Durham, NC

Karen Griffin, Vice Principal, Capitol Hill Cluster Schools

Judy Grumbacher, Fairfax County Public Schools, Fairfax, VA

Donna Hartney, HGL Associates, Arlington, VA

George Hein, Program Evaluation and Research Group, Lesley College, Cambridge, MA

Richard Hofmeister, Chief, Special Assignments and Photography Branch, Smithsonian Institution, Washington, D.C.

Chris Holle, East Los Angeles Science Center, Los Angeles, CA

Veola Jackson, Principal, Capitol Hill Cluster Schools

Ulysses Johnson, Vice Principal, Watkins School

Mary Ellen McCaffrey, Photographic Production Control, Smithsonian Institution

Stephanie Martin, Vice-President, Special Programs, Ohio Center of Science and Industry, Columbus, OH

Seliesa Pembleton, Naturalist, Hard Bargain Farm Environmental Center, Accokeek, MD

Dane Penland, Staff Photographer, Smithsonian Institution Photographic Services

Sabra Price, Program Evaluation and Research Group, Lesley College, Cambridge, MA

Marc Rothenberg, Editor, Joseph Henry Papers, Smithsonian Institution

David Savage, Montgomery County Public Schools, Rockville, MD

Blanche Wine, American University, Washington, D.C.

The NSRC would like to thank the following individuals and school systems for their assistance with the national field testing of the *Magnets and Motors* unit:

Elsa Aragon, Teacher, Stober Elementary School, Lakewood, CO

Sue Brandon, Science Coordinator, Jefferson County Public Schools, Golden, CO

Sarah Dargan, Teacher, Dilworth Elementary School, Charlotte, NC

JoAnn DeMaria, Teacher, Hutchinson Elementary School, Herndon, VA

Kevin Engbretson, Teacher, Dilworth Elementary School, Charlotte, NC

Rachida Faille, Teacher, Dilworth Elementary School, Charlotte, NC

Michael Howard, Science Coordinator, Fayette County Public Schools, Lexington, KY

Andrea Irby, Principal, Garrison Elementary School, Washington, DC

Margaret Jackson, Teacher, Garrison Elementary School, Washington, DC

Connie Jordan, Teacher, Arlington Elementary School, Lexington, KY

Bruce Laird, Teacher, Elk Creek Elementary School, Pine, CO

Joan Mazurek, Teacher, Breckinridge Elementary School, Lexington, KY

Joan Newman, Assistant Principal, Dilworth Elementary School, Charlotte, NC

Sandra Scott, Teacher, James Lane Allen Elementary School, Lexington, KY

Jan Withington, Teacher, South Lakewood Elementary School, Lakewood, CO

We also are indebted to the members of the STC Advisory Panel, who reviewed the unit and made suggestions for its improvement, and to George Hein and Sabra Price of Lesley College, who evaluated the unit.

We would like to thank Robert McC. Adams, Secretary of the Smithsonian Institution, and Frank Press, President of the National Academy of Sciences, for their vision and support in helping the NSRC to undertake this project.

Douglas Lapp
Executive Director
National Science Resources Center

STC Advisory Panel

Peter Afflerbach, Director, The Reading Clinic; Associate Professor, Curriculum and Instruction, University of Maryland, College Park, Maryland

David Babcock, Director, Board of Cooperative Educational Services, Second Supervisory District, Monroe-Orleans Counties, Spencerport, New York

Judi Backman, Math/Science Coordinator, Highline Public Schools, Seattle, Washington

Albert Baez, President, Vivamos Mejor/USA

Andrew R. Barron, Professor of Chemistry, Harvard University, Cambridge, Massachusetts

DeAnna Banks Beane, Project Director, YouthALIVE, Association of Science-Technology Centers, Washington, D.C.

Al Buccino, Education Advisor, Office of Science and Technology Policy, Executive Office of the White House, Washington, D.C.

Audrey Champagne, Professor of Chemistry and Education and Chair, Educational Theory and Practice, School of Education, State University of New York at Albany, Albany, New York

Sally Crissman, Faculty Member, Lower School, Shady Hill School, Cambridge, Massachusetts

Gregory Crosby, National Program Leader, U.S. Department of Agriculture Extension Service/4-H, Washington, D.C.

JoAnn E. DeMaria, Teacher, Hutchison Elementary School, Herndon, Virginia

Hubert M. Dyasi, Director, Workshop Center for Open Education, City College of New York, New York, New York

Timothy H. Goldsmith, Professor of Biology, Yale University, New Haven, Connecticut

Charles Hardy, Assistant Superintendent, Instruction and Curriculum, Highline Public Schools, Seattle, Washington

Patricia Jacobberger Jellison, Geologist, National Air and Space Museum, Smithsonian Institution, Washington, D.C.

Patricia Lauber, Author, Weston, Connecticut

John Layman, Professor of Physics, University of Maryland, College Park, Maryland

Sally Love, Museum Specialist, National Museum of Natural History, Smithsonian Institution, Washington, D.C.

Phyllis R. Marcuccio, Assistant Executive Director for Publications, National Science Teachers Association, Arlington, Virginia

Lynn Margulis, Professor of Biology, University of Massachusetts, Amherst, Massachusetts

Margo A. Mastropieri, Co-director, Mainstreaming Handicapped Students in Science Project, Purdue University, West Lafayette, Indiana

Richard McQueen, Specialist, Science Education, Multnomah Education Service District, Portland, Oregon

Alan Mehler, Professor, Department of Biochemistry and Molecular Science, College of Medicine, Howard University, Washington, D.C.

Philip Morrison, Professor of Physics, Emeritus, Massachusetts Institute of Technology, Cambridge, Massachusetts

Phylis Morrison, Educational Consultant, Cambridge, Massachusetts

Fran Nankin, Editor, *SuperScience Red*, Scholastic, Inc. New York, New York

Jerome Pine, Professor of Physics, California Institute of Technology, Pasadena, California

Harold Pratt, Director, Middle School Science Project, Jefferson County Public Schools, Golden, Colorado

Wayne E. Ransom, Executive Director of Educational Programs, Franklin Institute, Philadelphia, Pennsylvania

David Reuther, Senior Vice-President and Editor-in-Chief, William Morrow Books, New York, New York

Robert Ridky, Associate Professor of Geology, University of Maryland, College Park, Maryland

F. James Rutherford, Chief Education Officer and Director, Project 2061, American Association for the Advancement of Science, Washington, D.C.

David Savage, Training Specialist, Office of Instruction and Program Development, Montgomery County Public Schools, Rockville, Maryland

Thomas E. Scruggs, Co-director, Mainstreaming Handicapped Students in Science Project, Purdue University, West Lafayette, Indiana

Larry Small, Science/Health Coordinator, Schaumburg School District 54, Schaumburg, Illinois

Michelle Smith, Publications Coordinator, Office of Elementary and Secondary Education, Smithsonian Institution, Washington, D.C.

Susan Sprague, Director of Science and Social Studies, Mesa Public Schools, Mesa, Arizona

Arthur Sussman, Director, Far West Regional Consortium for Science and Mathematics, Far West Laboratory, San Francisco, California

Emma Walton, Science Program Coordinator, Anchorage School District, and Past President, National Science Supervisors Association, Anchorage, Alaska

Paul Williams, Director, Center for Biology Education; and Professor, Department of Plant Pathology, University of Wisconsin, Madison, Wisconsin

Contents

	Foreword	iii
	Acknowledgments	iv
	Contents	vii
	Unit Overview and Materials List	1
	Teaching Strategies and Classroom Management Tips	3
Lesson 1	Getting Started—Pre-Unit Assessment	7
Lesson 2	What Can Magnets Do?	11
Lesson 3	How Can You Find Out What Magnets Can Do?	15
Lesson 4	Measuring Magnets	21
Lesson 5	Building a Compass	27
Lesson 6	Using a Compass: Which Way Is Which?	37
Lesson 7	Creating Magnetism through Electricity	43
Lesson 8	Making Magnets with Electricity	49
Lesson 9	Designing an Experiment to Test the Strength of an Electromagnet	57
Lesson 10	Testing an Electromagnet	67
Lesson 11	Showing Others What You Have Learned	71
Lesson 12	Making a Motor	77
Lesson 13	Building a Spinning Coil Motor	83
Lesson 14	What Is Inside an Electric Motor?	89
Lesson 15	How Does a Motor Work?	95
Lesson 16	Generating Electricity	99
Appendix A	Post-Unit Assessments	103
Appendix B	Teacher's Record Chart of Student Progress	109
Appendix C	Making a Mystery Box	111
Appendix D	Background: Electric Circuits	115
Appendix E	Using a Wire Stripper	119
Appendix F	Making and Repairing Alligator Leads	123
Appendix G	Setting Up a Learning Center for *Magnets and Motors*	125
Appendix H	Bibliography	131
Appendix I	Materials Reorder List	135

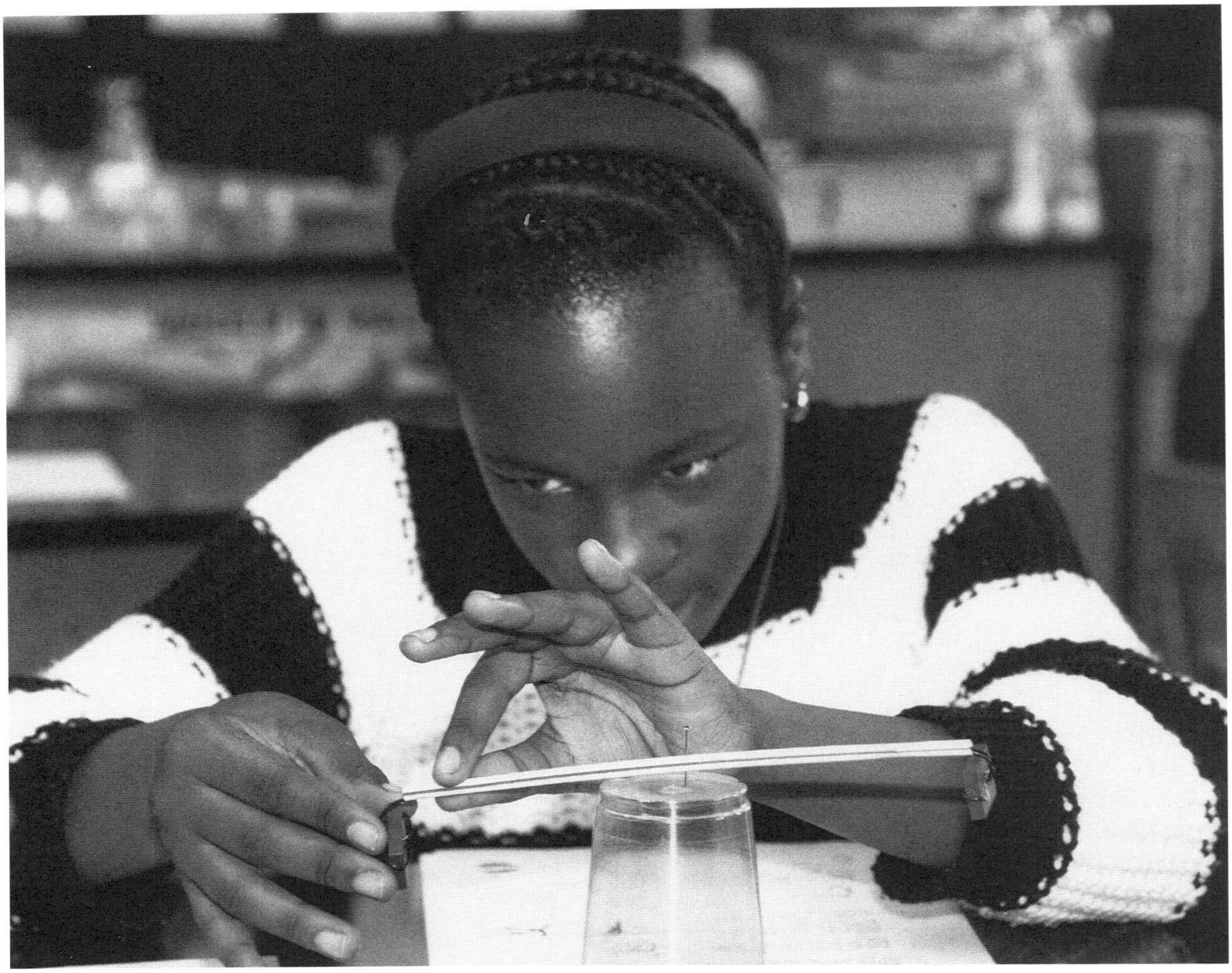

Unit Overview and Materials List

What makes your refrigerator door stay closed? That seems like a silly question, but we do take magnets for granted. There are many other important uses for magnets—telephones, televisions and video recorders, loudspeakers, electric motors to run everything from the largest to the smallest electric appliance, and even gasoline-powered automobiles (the starter motor, alternator, windshield wiper motor, and transformer all rely on magnetism!). The world would be a very different place if magnetism—and the electricity it allows us to produce—had never been investigated scientifically.

Magnets and Motors is a unit of 16 lessons about magnetism and electricity designed for 6th graders. It mirrors the historical development of our understanding and use of magnetism, electricity, and electromagnetism. The unit progresses through these phenomena in the same order that people first learned about them—magnets and compasses, electricity from batteries, then electromagnetism (electromagnets, motors, and generators). Opportunities are taken to integrate the science activities with other subjects such as language arts, mathematics, history, and geography.

Lesson 1 provides opportunities for students to take part in a preassessment brainstorming session and write and talk about what they know already about magnets and motors. Students also become familiar with the *Magnets and Motors Student Activity Book* and with keeping a student notebook.

In Lessons 2 through 6, students experiment with magnets and with a compass that they have made. They discover (or rediscover) the properties of magnets, learn a simple problem-solving technique, manipulate materials to build a working device, and test their own hypotheses. They also develop and practice language, classification, and manipulative skills.

Students observe and investigate magnetism's connection with electricity in Lessons 7 through 11. They explore the characteristics of a switch, simple circuits, and electromagnets, then work in teams to design and conduct an experiment with electromagnets and report their findings to their peers. In the process, they learn to evaluate alternatives, work with others, and present their ideas to a group.

In Lessons 12 through 15, students experiment with three different electric motors, including one they make. They practice the troubleshooting and manipulative skills acquired in earlier lessons and observe phenomena that will help them to make connections between electricity and magnetism. Based on their own experiences, they develop an understanding of how a motor works.

Lesson 16 enables students to learn how to produce electricity with an electric generator (a motor used "backwards"). With the electricity they generate, students light a bulb and make a motor turn.

This is an exciting unit for students. They will be quite energized while working with it, but also occasionally confused. A certain amount of anxiety is useful at times; it can cause students to become actively engaged in what they are doing and to come up with creative questions. Some of their questions will be puzzlers; after all, electricity and magnetism are puzzling phenomena which scientists continue to study. The teacher will need to gauge carefully how to respond to these questions. Often, the most effective response is to encourage children to find out more for themselves through additional experimentation or by seeking out additional information in science trade books.

Materials List

Below is a list of the materials needed for the *Magnets and Motors* unit.

- 1 Teacher's Guide
- 15 Student Activity Books
- 90 flexible magnets, 25 x 20 x 5 mm (1" x ¾" x 3/16") with a 5-mm (3/16-inch) hole
- 40 plastic cups
- 40 plastic lids
- 1 spool of light string
- 2 pencils, No. 2
- 30 wooden sticks, 15 cm x 4 mm (6" x 1/8") or 30 long toothpicks
- 15 tongue depressors, or wide "craft sticks"
- 15 boxes, cardboard, 10 x 5 x 20 cm (4" x 2" x 8")
- 10 packages of assorted objects, each containing:
  - steel washer
  - aluminum screen
  - brass fastener
  - rubber band
  - copper wire
  - recording tape
  - aluminum wire
  - steel nail
  - aluminum foil
  - brass washer
  - paper clip
  - pipe cleaner
  - golf tee
  - twist-tie
- 20 jumbo paper clips
- 150 No. 1 paper clips
- 500 steel washers, USS standard No. 10
- 30 magnetic compasses
- 30 straight pins, 2.5 cm (1 inch)
- 90 plastic drinking straws
- 120 single-color stickers
- 30 D-cell batteries
- 30 battery holders
- 30 bulbs
- 60 bulb sockets
- 1 wire-stripper tool
- 2 aluminum nails, 12D or larger
- 30 nails, common 12D
- 2 brass bolts, 8 cm x 6 mm (3" x ¼")
- 30 steel bolts, 12 mm x 8 mm (½" x 3/8")
- 30 steel bolts, 8 cm x 6 mm (3" x ¼")
- 2 steel bolts, 3.5 cm x 6 mm (1¼" x ¼")
- 2 steel bolts, 8 cm x 4 mm (3" x 1/8")
- 2 steel bolts, 8 cm x 8 mm (3" x 3/8")
- 2 steel bolts, 8 cm x 12 mm (3" x ½")
- 2 steel bolts, 10 cm x 6 mm (4" x ¼")
- 2 steel bolts, 13 cm x 6 mm (5" x ¼")
- 10 graph grid transparencies
- 8 transparency marker pens
- 10 envelopes
- 16 sheets of contruction paper, including:
  - 4 sheets pink
  - 4 sheets blue
  - 8 sheets yellow
- 60 alligator clips
- 1 roll (23 m, 50 feet) #20 bare wire
- 2 rolls (7.5 m, 25 feet) #20 coated hook-up wire, 1 roll each of two different colors
- 3 rolls (30 m, 100 feet) #22 coated hook-up wire (90 m)
- 1 roll (23 m, 75 feet) #28 enameled wire
- 30 sandpaper squares, 5-cm square (2-inches square)
- 6 screwdrivers
- 30 small, electric motors, Mabuchi RE-260, with wire leads (30 cm, 12 inch)
- 1 roll of PVC electrical tape, to make 30 double wires
- 60 rubber bands, No. 16
- 60 rubber bands, No. 84 or larger
- *30 student notebooks
- * Several sheets of newsprint
- * Overhead projector
- * Cellophane tape
- * Glue stick or paste
- * Large world map or globe

*Note: These items are not included in the kit. Including them would increase material and shipping costs, and they are commonly available in most schools or can be brought from home.

Teaching Strategies and Classroom Management Tips

The teaching strategies and classroom management tips in this section will help you give students the guidance they need to make the most of the hands-on experiences in this unit. These strategies and tips are based on the assumption that students already have formed many ideas about how the world works and that useful learning results when they have the opportunity to re-evaluate their ideas as they engage in new experiences and encounter the ideas of others.

Classroom Discussion: Class discussions, effectively led by the teacher, are important vehicles for science learning. Research shows that the way questions are asked, as well as the time allowed for responses, can contribute to the quality of the discussion.

When you ask questions, think about what you want to achieve in the ensuing discussion. For example, open-ended questions, for which there is no one right answer, will encourage students to give creative and thoughtful answers. Other types of questions can be used to encourage students to see specific relationships and contrasts or to help them to summarize and draw conclusions. It is good practice to mix these questions. It also is good practice always to give students "wait-time" to answer. (Some researchers recommend a minimum of 3 seconds.) This will encourage broader participation and more thoughtful answers.

Brainstorming: A brainstorming session is a whole-class exercise in which students contribute their thoughts about a particular idea or problem. It can be a stimulating and productive exercise when used to introduce a new science topic. It is also a useful and efficient way for the teacher to find out what students know and think about a topic. As students learn the rules for brainstorming, they will become more and more adept in their participation.

To begin a session, define for students the topics about which ideas will be shared. Tell students the following rules:

- Accept all ideas without judgment.
- Do not criticize or make unnecessary comments about the contributions of others.
- Try to connect your ideas to the ideas of others.

Ways to Group Students: One of the best ways to teach hands-on science lessons is to arrange students in small groups of two to four. There are several advantages to this organization. It offers pupils a chance to learn from one another by sharing ideas, discoveries, and skills. With coaching, students can develop important, interpersonal skills that will serve them well in all aspects of life. Finally, by having children help each other in groups, you will have more time to work with those students who need the most help.

As students work, often it will be productive for them to talk about what they are doing, resulting in a steady hum of conversation. If you or others in the school are accustomed to a completely quiet room at all times, this new, busy atmosphere may require some adjustment. It will be important, of course, to establish some limits to keep the noise under control.

Safety: You will want to warn your class not to experiment with electric outlets and appliances at home or in school. The high voltage supplied from these outlets can deliver a fatal shock. Also:

- D-cell batteries (the kind supplied in this unit) will not give even a mild shock unless more than two dozen are connected in series. However, these batteries can generate significant amounts of heat if a wire is connected to both ends, creating a short circuit. Therefore, caution students to avoid connecting short circuits for long periods of time.

- Do not use rechargeable batteries. There have been reports of very hot wires when these batteries are short-circuited.

- If a light bulb breaks, there probably will be broken glass on the floor. You may find it helpful to establish a system for cleanup under your supervision.

Handling Materials: To save time and reduce disruption, you may want to set up a system for storing and distributing materials that works for you and your class. Some suggestions are given below. Additional tips are found throughout the unit where you see this icon:

- A "cafeteria-style" distribution station has proven to be a time-saver for teachers. Select one large area or several small areas of the room where students can walk by in single file on both sides of the supplies. Line up all of the materials on a table or group of desks and place a printed label on each item telling students how many to take.

- Plastic cups are used many times in this unit to support various apparatuses. These cups (and matching lids) also are effective containers in which students can gather and store materials.

- A complete list of the materials needed for this unit is included at the end of the **Unit Overview**. The insulated wire must have the ends stripped so that it can be used in the circuits. Therefore, you may want to select several students to serve as wire-stripper helpers throughout the unit. (For instructions on stripping wire, please see **Appendix E**.)

- Learning how to make and repair alligator clip leads also will be important—for you and your students. See **Appendix F**.

- You may find it useful to preview each lesson ahead of time. Some have specific suggestions for handling the materials needed that day.

Setting Up a Science Learning Center: Supplemental science materials should be given a permanent home in the classroom in a spot designated as the science learning center. Such a center can be used by students in a number of ways: as an "on their own" project center, a trade-book reading nook, or simply a place to spend unscheduled time when assignments are done.

In order to keep interest in the center high, it is a good idea to add to it frequently. **Appendix G** contains specific suggestions and illustrations for such a center.

Curriculum Integration: There are many opportunities for curriculum integration in this unit. Look for the following icons for math, reading, writing, art, speaking, and social studies that highlight these opportunities.

Evaluation: In the STC project, evaluation tools are included throughout each unit, and post-unit assessments are provided at the end. This arrangement is intended to help you assess what students know and monitor how they are progressing, making it easier for you to provide assistance to students who need it, to go over materials students did not grasp, and to report to parents on student progress. The assessments provided also are intended to be directly helpful to students, giving them an opportunity to reflect on their own learning, gain confidence by viewing their own progress, articulate the ways in which they want and need to grow, and formulate further questions.

Evaluation Preparation: To facilitate successful documentation and assessment of students' learning in the specific content areas of a unit as well as in the development of relevant skills, you should be prepared to:

- **Collect pre/post information.** One of the best indicators of student learning comes through gathering information from an identical activity or discussion conducted both before and after a unit. The pre-test suggested for use in Lesson 1 of this unit also is a suggested post-test. Note that comparison of non-identical materials generated early and late in a unit also can help you gauge growth in learning. You will want to make sure that pretest and post-test work is dated.

- **Encourage students to use notebooks in a way that is useful to you and to them.** Student notebooks provide information about student progress. To ensure that you will be able to make the best use of these notebooks, be sure to ask students to:
  - make only one entry per page;
  - date each entry;
  - write out conclusions and interpretations of experiments;
  - write explanations to charts, tables, and graphs; and
  - include the question when writing out answers.

  Student notebooks are a particularly handy and effective way to share student progress and accomplishments with parents and other interested adults as well as with the students themselves.

- **Save other student work.** Other student work, such as the compasses they create, the motors they build, and the methods they invent to generate electricity, provide concrete demonstrations of student knowledge.

- **Observe.** Invaluable for assessment are your ongoing observations of students as they work, written in your own notebook or on file cards.

- **Allow time for oral presentations.** Oral presentations by students can be useful vehicles for assessment.

A variety of post-unit assessment instruments is offered in **Appendix A**. From those suggested, you will want to choose only the instruments that are most appropriate for measuring the achievement level of your class. Consider using more than one, in order to give students with differing learning styles a chance to express their knowledge and skills. Different styles of assessment have been shown to be particularly helpful in increasing the precision of the assessment of girls and minorities—two groups that have historically underperformed in science.

Please note that **Appendix B** includes a black line master for a "Teacher's Record Chart of Student Progress." You may want to reproduce this chart at the beginning of the unit to help you record individual student achievement throughout the unit. Please remember that most students at the grade level targeted by this unit will not be able to master and articulate the full list of skills.

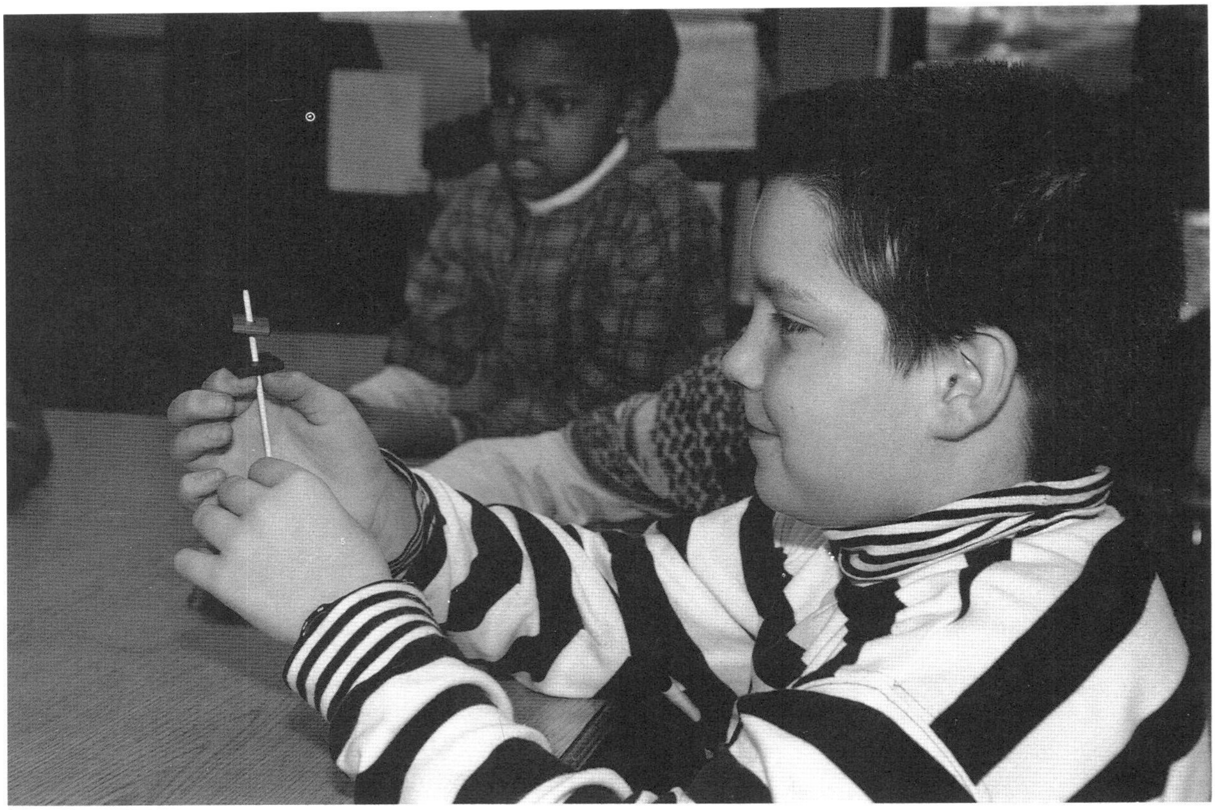

LESSON 1

Getting Started—Pre-Unit Assessment

Overview

This lesson consists of a series of pre-unit assessment exercises. Students discuss what they already know about magnets and motors, and they learn to record their thoughts and activities in a student notebook or journal. You can use the information gathered in this lesson as a way to tailor the unit to the needs of the class. You and your students also will be able to use this information to gauge what students have learned as they progress.

Objectives

- Students express their preconceptions about magnets and motors.
- Students learn how to set up and keep records in a student notebook.

Background

Throughout this unit, students will be encountering phenomena and formulating questions about their experiences—some of the processes of science. As children learn in this way, it is helpful for them to be able to reflect on what they have learned. A student notebook is an aid to reflection; it provides students with a record of their questions, ideas, and experiences.

It also is necessary for the teacher to be aware of the progress that each student has made. The student notebook is an important part of the portfolio of students' work that you can use to assess their learning. The pre-test suggested in this lesson, when coupled with a similar or identical post-test at the end of the unit, also can provide information on how much students' thinking has developed.

Finally, a brainstorming session (described on pg. 3) enables students to express what they know already about magnets and motors. This allows you to tailor the lessons in this unit to a particular class of students. It also provides yet another baseline for later assessment of their progress.

Materials

For each student
1   student notebook
1   copy of **Activity Sheet 1, What Do You Already Know about Magnets and Motors?**

For the class
Several sheets of newsprint

LESSON 1

Figure 1-1

Preparation

1. Prepare one copy of **Activity Sheet 1** for each student.
2. Arrange the newsprint so that it can be used for a brainstorming session.

Procedure

1. In order to get an idea of what the students know about magnets, electricity, and motors before beginning this unit, you probably will want to start the lesson by explaining that a pre-assessment is a way of measuring how much students have grown. It is not a test, and there are no right or wrong answers. It is a good idea to let students know that it is all right to write things such as, "I really don't have any idea about this at all," as long as that is a true statement. You may want to compare a pre-unit assessment with the marks that people sometimes make on walls to record how tall they are at various ages.
2. Let the students work on **Activity Sheet 1.**
3. After they have had enough time with the **Activity Sheet**, explain to them what a brainstorming session is and the rules that make it a useful way to discuss ideas about a new topic (see pg. 3). You might want to list these rules in the classroom.
4. Next, ask students to share with the class what they know about magnets and motors. You may want to make one list for magnets and another list for motors, writing them on the newsprint for later reference.

Final Activities

1. Ask the students to think of the questions they have about magnets and motors and then to write the questions they have in their notebooks. Discuss with them the reasons for recording their questions in a student notebook. Ask them what other kinds of information it might be useful to record.
2. Ask the students to share some of their questions with the class in order to help you make a list on a sheet of newsprint. Urge students to suggest how they might find answers to their questions.

LESSON 1

3. Introduce the students to the Student Activity Book. Explain that the book does not contain answers—only suggestions and questions that might be interesting to explore. This may be a good time to talk to students about questions that do not have one short, simple, "right" answer, such as, "What can magnets do?" and "How do you know what magnets can do?"

4. Remind the students to write in their notebooks and **not** in their Student Activity Books.

Evaluation

1. **Activity Sheet 1** and the lists generated during brainstorming can be used as a basis for assessing what children are learning throughout the unit.

2. The notebooks students establish today will become a valuable evaluation tool. They will serve as portfolios of student work containing drawings, ideas, discoveries, and questions. In most cases, they will show gradual progress in the development of these skills.

3. Use the checklist provided in **Appendix B** to note your students' progress.

LESSON 1

What Do You Already Know about Magnets and Motors? **Activity Sheet 1**

NAME: _____

DATE: _____

1. Think of all the places that electricity comes from. Make a list of at least five of those places below.

2. Draw a picture of what you think an electric motor looks like. Please put labels on your drawing in order to tell other people about it.

3. What questions do you have about magnets, electricity, and motors that you would like to investigate?

LESSON 2

What Can Magnets Do?

Overview

In this lesson, students will discover several properties of and uses for magnets. They also will have an opportunity to generate questions and to voice the ideas they have about magnets.

Objectives

- Students learn several properties of and uses for magnets.
- The teacher learns the ideas students have about magnets.

Background

Magnetism is one of the most interesting—and useful—mysteries of the natural world. Existing within the earth itself, magnetism can levitate objects, act through solid objects (including human hands), and pull on things, without touching them. Magnetism is what makes electric motors spin, loudspeakers and earphones work, refrigerator doors stay closed, and televisions, video recorders, and all household electricity (produced by electric generators, which use magnets) possible.

Although permanent magnets can be manufactured in many different forms, they all share a common characteristic: they must contain either iron, or nickel, or cobalt, or a combination of these elements. Not all metals are attracted to a magnet—a fact that often confuses students. For example, aluminum foil is not magnetic, but it looks somewhat like steel, which is magnetic.

Flexible magnets, like those used in this unit, are made primarily of rubber or plastic, but they also contain many tiny particles of iron or **ferrite**, a class of materials containing iron. Similarly, magnetic recording tape (tape recorder tape and videotape) is made of plastic that has been coated with a very thin layer of an iron compound.

Some students will want to know why only certain metals are attracted to a magnet and why some seemingly nonmetallic objects also are attracted. You can take advantage of difficult questions like these to challenge students by asking them: "How can we find out?"

Management Tip: Keep magnets at least 30 centimeters (12 inches) away from computer disks and audio tapes. The information on them can be erased by magnets.

LESSON 2

Materials

For each student
- 1 student notebook
- 2 flexible magnets, 25 x 20 x 5 mm (1" x ¾" x 3⁄16"), with a 5-mm (3⁄16-inch) hole in the center
- 1 wooden stick, 15 cm x 4 mm (6" x ⅛"), or a large wooden toothpick, approximately the same size
- 1 piece of string, 30 cm (12 inches) long
- 1 plastic cup and lid

For the class
- 1 set of demonstration materials:
  - 1 small, electric motor with wire leads (like the Mabuchi RE-260 model suggested in the materials list)
  - 1 plastic drinking straw
  - 1 battery and battery holder
  - OR
  - 1 student- or teacher-supplied motorized toy, with batteries
  - OR
  - 1 shoe box and 2 strong magnets

Preparation

1. Cut one piece of light string 30 cm (12 inches) long and one 15-cm (6-inch) long wooden stick for each student.
2. Designate a place in the room for students to pick up and return their materials. Arrange the materials for collection. (Please see the suggestions on pg. 4.)
3. Prepare and rehearse one or two of the demonstrations listed in Step 1 of the **Procedure** section. Try to generate as much excitement and mystery as possible about what is inside the box or motor.

   If you are dismantling a motorized toy, take apart as much of it as possible before class (take out all of the screws except for one and hold it together with rubber bands, for example).

Procedure

1. To pique your students' interest, perform one or two of the following demonstrations:
   - Take apart a motorized toy, and demonstrate that paper clips stick to the small motor inside.

12 / What Can Magnets Do?

LESSON 2

- Tape two strong magnets inside a sealed shoe box and demonstrate that certain objects (scissors, stapler) stick to the box for some unknown reason. Tell students that they will never be allowed to look inside the box.

- Attach a straw "propeller" to an electric motor and run it for the students. Demonstrate that paper clips stick to the motor.

The idea is, first, to get the students' attention and, second, to get them to begin wondering about what might be causing things to stick to the motor or box. Students' responses will range from "velcro" to "tape" to "magnets." Respond by saying, "Maybe—how could you find out?" Most students probably will suggest opening up the motor or box and looking inside. Tell them that they will get the opportunity to open up a motor in a later lesson, but initially, you want them to experiment with one of the things they think might be in the box or motor—magnets.

You may want to reinforce the experience by discussing with students the ways in which scientists sometimes find out things without actually seeing them with their eyes—by looking for other evidence (see No. 1 in **Extensions**).

2. Explain to students the procedures for distributing and returning materials.

3. Distribute the materials to the students. Ask them to work with the materials to try and figure out for themselves "what magnets can do" (the Student Activity Book also asks them to do this). Observe what the students do with the materials. They will want to show you what they found out.

4. After about 10 or 15 minutes, ask the students to return the materials. Invite them to tell the class what they discovered about magnets. Have them help you to make a list of "magnet characteristics" on the board.

5. Lead a discussion of the possible uses of magnets. Generate a list on the board. Questions that may help to get the discussion rolling include:

    - How could magnets be used to hold a door closed?
    - Are magnets used in anything that you play with?

6. Ask the students to draw pictures in their student notebooks and to label them to illustrate what magnets can do. Tell them to save these drawings; they will refer to them later in the unit.

Final Activities

Give the students additional time to write in their notebooks. Suggest that they include what they now know about magnets and what they wonder about when they think of magnets. Remind them that, later, they will be able to look back at their notebooks and remember some of their experiments and what they have learned. Remind them to write the date on their work.

Extensions

1. Here are some ideas for beginning a discussion that could be held after demonstrating that certain objects stick to the shoe box and/or that paper clips stick to motors. The point of the discussion would be to make it clear to the students that scientists often must solve problems about phenomena that they cannot see with their eyes or experience firsthand. They must use other kinds of evidence to piece together what is going on.

What Can Magnets Do? / **13**

LESSON 2

For example, to develop theories about what is deep inside the earth, scientists examine how earthquake waves move through it, how continents drift, and how volcanoes erupt. Similarly, no one has ever "seen" a "black hole" in space, yet there is strong evidence that they exist (high-energy electromagnetic waves, or X rays, coming from a place where no object is seen).

There are other examples. Pluto's atmosphere, the way that birds navigate, and the effect of a drug on a disease are all areas that scientists explore with other than visual evidence.

2. Ask the students to draw an imaginary machine or a system of machines that uses magnets to help them perform a common task, such as getting ready for school or cleaning their room.

Evaluation

1. Observe the students while they are exploring what magnets can do. Notice how they use the materials. Do they focus on their own ideas or do they usually imitate what other students have discovered? Do they talk about what they think is happening?

2. The pictures that students draw in their notebooks will show what they are thinking about magnets. Also, some students may use terms like "south magnetic pole" or "magnetic field" (see Lesson 5 for background) to label their drawings. Others still may be confused about the difference between "push" and "pull." Strange ideas can result from students trying to reconcile their own experiences with magnets with what they have been told about them.

3. Save the list that you and the class make of the characteristics of magnets. Your observations, along with the products of the class and the individual students, will give you a baseline against which to compare their future behavior and help you assess the progress of individual students.

| LESSON 3 | # How Can You Find Out What Magnets Can Do? |

Overview

Now that the students have had experience with what magnets can do, they are asked to use what they know about magnetism to make predictions of the behavior of several different objects. The purpose of the predictions is for the students to see what they are thinking—to examine their own ideas.

Objectives

- Students learn to make and test their own hypotheses.
- Students become proficient at classifying objects and organizing information.
- Students observe the characteristics that distinguish a magnet from other magnetic materials.

Background

Somewhere along the line, many students get the idea that learning science means knowing the "right" answers to questions about the natural world. But in science, new questions are being asked all the time and answers sought, often because of new observations. It is an exciting and fruitful process. Over the course of this unit, try to emphasize to students that learning science is not so much a process of being right or wrong as it is of finding out new things. The goal is to help them open their minds and prevent them from slipping into the "how-many-did-you-get-right?" syndrome.

As students make hypotheses—tentative assumptions for the purpose of testing the validity of an idea—about the magnetic characteristics of a variety of objects and materials, many will think that aluminum foil and copper wire are magnetic. They will predict that these materials will be attracted to a magnet because they are metal. Students will be surprised to discover that not all metals are magnetic—and perhaps even embarrassed by the fact that they "guessed wrong." Here is an opportunity to downplay their embarrassment and to emphasize the importance of learning something new. (Some students are so possessed by the desire to be "right" that they will write their predictions in pencil and then change them if the experiment disagrees with their predictions.)

It also is important for students to discover at this point that copper and aluminum are not magnetic. In Lesson 7, they will learn that electricity

LESSON 3

moving through these materials will make them temporarily magnetic. The magnetic effect of electric current will be easier for students to comprehend if they first learn that not all metal wires are magnetic. Students may point out that the wires in twist-ties and pipe cleaners **are** magnetic. Asking questions like "Are all metals the same?" and "What are some ways that you can tell different metals apart?" will help students to investigate and clarify this distinction.

Materials

For every student
- 1 student notebook

For every two students
- 1 flexible magnet, 25 x 20 x 5 mm (1" x ¾" x ³⁄₁₆"), with a 5-mm (³⁄₁₆-inch) hole in the center
- 1 mystery box

For every four students
- 1 package containing assorted materials:
    - Aluminum foil
    - Aluminum screen
    - Aluminum wire
    - Brass fastener
    - Brass washer
    - Copper wire
    - Golf tee
    - Paper clip
    - Pipe cleaner
    - Recording tape
    - A piece of rubber band
    - Steel nail
    - Steel washer
    - Twist-tie

Preparation

1. Provide one package of assorted materials for every four students (two pairs will share).
2. See **Appendix C** for instructions to assist you or helpers in constructing enough mystery boxes for every two students to share. Alternatively, students could construct the boxes as part of this lesson, using the **Appendix C** instructions. See Step 9 of **Procedure**.

Procedure

1. Ask students how they would decide whether something is magnetic or not. Several students will probably say that they would be able to see if the objects were attracted to a magnet or not. Ask how they would be able to "see." Ask them to suggest ways to tell if an object is only very slightly attracted to a magnet.
2. Next, ask them to discuss how they would distinguish between a magnet (an object that attracts and repels another magnet) and something that

16 / How Can You Find Out What Magnets Can Do?

Figure 3-1

Inside a mystery box

A mystery box ready for use

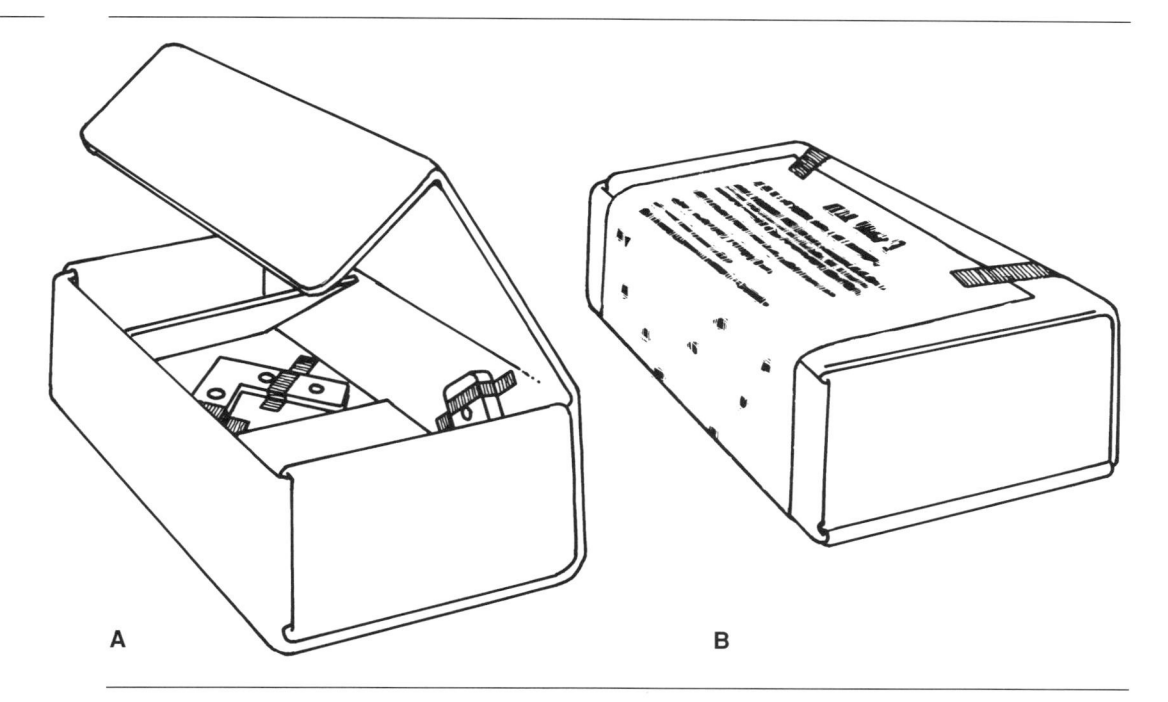

A

B

is made of a magnetic material (an object that is attracted to a magnet, but not repelled). Pay close attention to the students' responses and evaluate whether they have learned that two magnets can attract and repel one another, while a magnet and magnetic material can only attract one another. Demonstrate this difference by using two magnets first, then a magnet and a paper clip, if you think it will help students to understand.

3. Discuss the difference between the words "push" and "pull" and how they relate to the words "attract" and "repel."

4. Show students the package of materials that they are going to test for magnetic properties. Tell them that two pairs of students—teams of four students—will share one package of materials. Explain that they will be working with a lot of information and materials.

5. Ask students for their ideas about how best to organize the information in their notebooks. Let them guide you through the production of a sample data table on the chalkboard before they set up their own data tables in their student notebooks. Students may need to be reminded of the importance of a title and columns. The title and column headings could be:

Results of Testing for Magnetic Properties

Description of Item	What I Think Will Happen	Why I Think It Will Happen	What I Found Out by Testing

Demonstrate for students how to enter a few of the items on the chart. For example, you could start with "wooden golf tee," then predict "Not magnetic because it is wood." After testing, you could enter: "It was not attracted to the magnet."

How Can You Find Out What Magnets Can Do? / 17

LESSON 3

6. Distribute the package of materials to each team of four students. Once the teams have their materials, they can begin to list each object to be tested and their predictions of whether it is magnetic or not. Remind the students (as the Student Activity Book does) that it is more important for them to be aware of what they are learning than for them to worry about making correct predictions. Tell students they will get magnets after they have completed their listing and predicting.

7. Give each pair of students a flexible magnet after they have completed their predictions and their reasons. Tell them to begin testing the materials one at a time. Observe how the students interact with the materials and with their partner. Some students may need to be encouraged to try testing the materials themselves (instead of letting their partners do it all).

8. After students have finished testing the materials, ask them to look back at their predictions and reasons to see if any of their tests did not turn out as they predicted.

9. The pairs of students will finish testing the materials at different times. Give a mystery box to those who finish early and ask them to investigate its properties. Alternatively, you may want to ask pairs of students to construct a mystery box to exchange with other teams. **Appendix C** contains suggestions for its contents and instructions for construction.

Final Activities

After all of the groups have finished, ask the students to tell the class what they found out. This will help you to evaluate what they are able to generalize from their findings. You may need to guide the discussion toward conclusions that not all metals are magnetic, that there is a difference between magnetic materials and magnets, and that some objects that are nonmetallic contain a magnetic metal. Ask questions to get all of the students involved in the discussion. Some possible questions are:

- What similarities do you notice among the magnetic objects?
- What do you think makes an object magnetic?
- How can you use magnets to help find out what certain things are made of?
- Why would you want some things to be nonmagnetic?
- What ways can you find to make something a magnet?
- What do you think would happen if everything were magnetic?

Extensions

1. Challenge the students to use magnets to help them find out what objects are made of—in the classroom or at home.

Safety Reminder

Warn students that electricity from wall sockets is very dangerous and should be left alone.

---

18 / How Can You Find Out What Magnets Can Do?

2. Challenge the students to find things in their home that use magnets. Ask them to keep a record in their notebooks of what they find.

3. This lesson provides an opportunity for students to think about the names objects have and the reasons for naming things. You may want to discuss with students the possible origins of words such as "washer," "pipe cleaner," "tee," "foil," "twist-tie," "paper clip," and others. Which words are descriptive? What else could these items be called?

4. The various materials in the assortment package present an identification challenge for students. How do you know that the paper fastener is made of brass? What is brass? Where does steel come from? How are paper clips made? You may want to invite students to write what they suppose to be true about a related topic, then to do some independent research and write an account of what they found out.

5. Ask students whether they think a U.S. nickel is magnetic or not. Ask them to find out what a nickel is made of. Nickels are currently 75 percent copper and 25 percent nickel so they are only very slightly magnetic. One way to demonstrate this is to suspend a nickel from a thread with tape and observe the effect that a strong magnet has on it. The effect is *very* slight. You may want to discuss the composition or origin of coins after this. Ask students if they can think of any problems that strongly magnetic coins might cause for vending machine manufacturers.

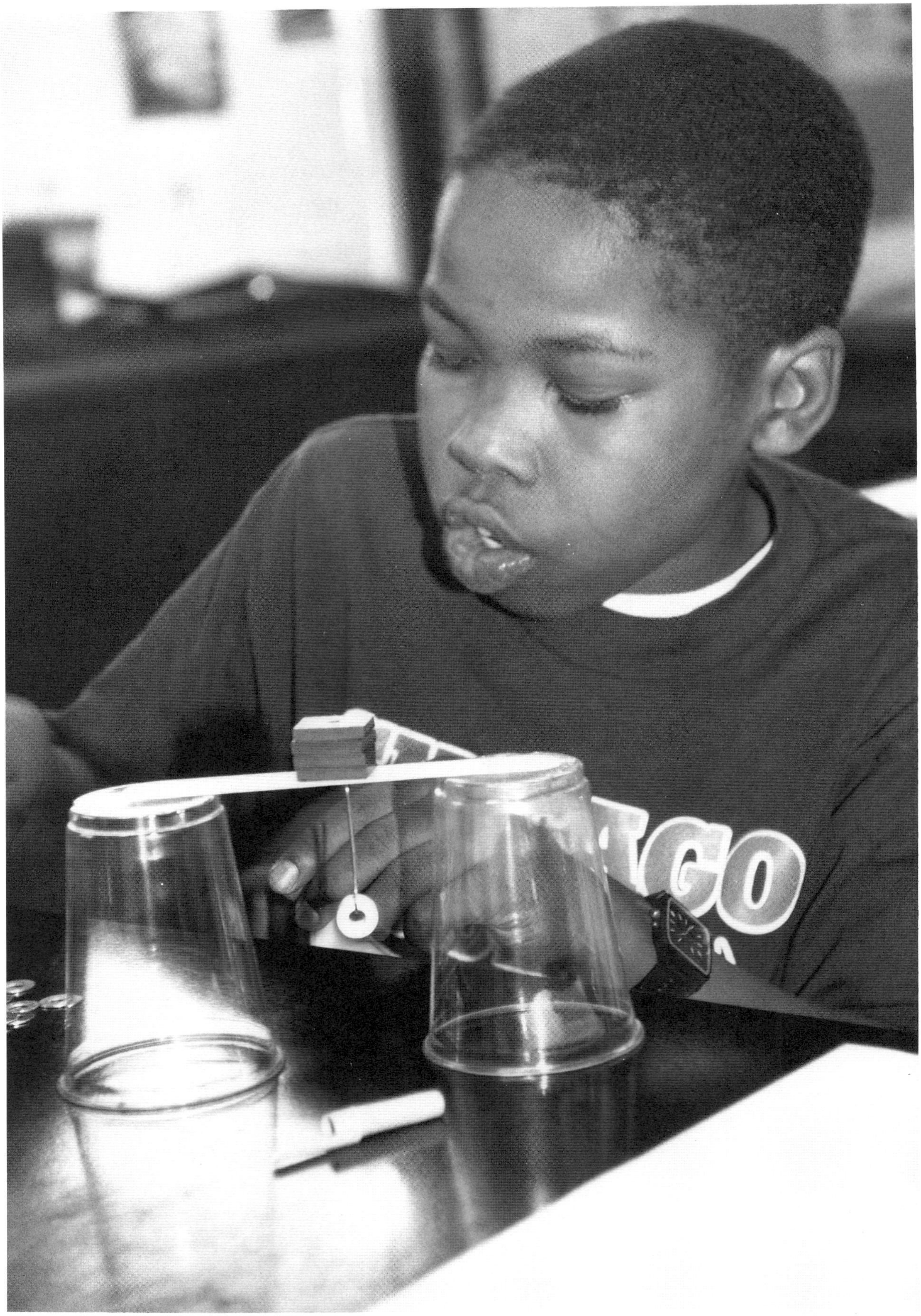

| LESSON 4 | **Measuring Magnets** |

Overview

In Lessons 2 and 3, students had a variety of experiences with magnets. Now, in this lesson, they will learn about "fair tests" and, in teams, conduct an experiment to find out how strong different combinations of magnets are. Then the results of the experiment will be graphed. A similar experiment using electromagnets will be designed by students in Lesson 9 and performed in Lesson 10.

Objectives

- Students begin to understand the concept of a controlled experiment, or a "fair test."
- Students develop skill in manipulating materials.
- Students become skillful in conducting an experiment in a systematic way.
- Students learn one way of varying magnetic strength.
- Students use graphs to communicate their results.

Background

Magnets are used in many, simple ways—to hold cabinet or refrigerator doors closed, for example. The magnet that holds a door closed must be strong, but not too strong. How strong does the magnet need to be? How strong is too strong? Obviously, someone had to do some experiments to find the answers to these questions.

A "fair test" is another way of doing a "controlled experiment." Students have ideas about what is fair and what is unfair. They know, for example, that a fair race is one run with all runners on level ground, no wind advantage, and no head starts. Similarly, a "fair test" is an experiment that keeps all things the same, except the thing that you are trying to investigate.

The experiment that the students perform in this lesson explores how the strength of a magnet changes when it is combined with other magnets. The variables that need to be controlled in this experiment are the kind of magnets used (it would not be fair to use magnets of different strengths), the thickness of the tongue depressor (so that the same distance from the magnet is maintained), and the weight of the washers (it would not be fair if one washer weighed more than the others). Figure 4-1 shows the experimental setup.

LESSON 4

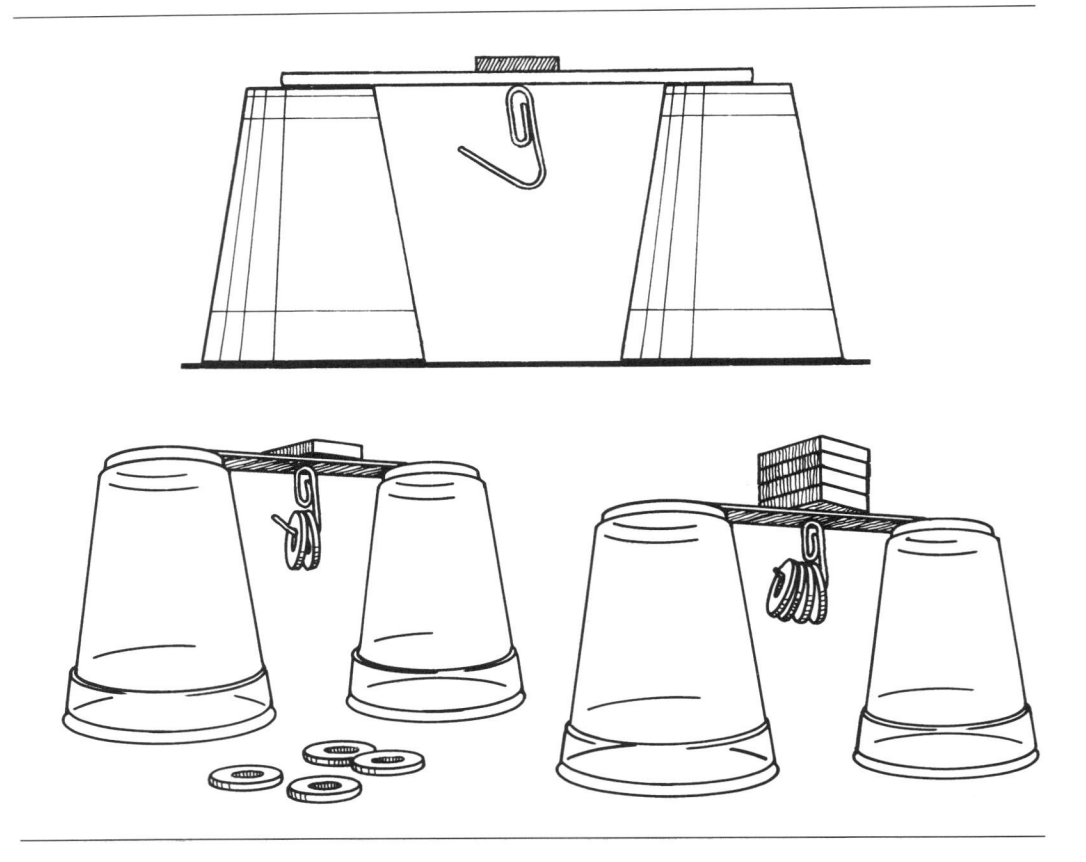

Before the students begin the experiment, you will be asking them to think about whether two magnets are stronger than one magnet. Do the magnetic strengths of two magnets combine? The answer will not be obvious until the experiment is performed.

You also will be asking the students how the test can be made fair. Asking them for their ideas will encourage the students to think about what they are doing. You may want to give them an example of an unfair test such as the following.

Imagine two students (pick any two) who wanted to know which of them had a stronger left arm. They decided to arm wrestle, but the first student used the whole hand and the second student used only the little finger. Was this a "fair test"? How could you make it fair?

Materials

For each student

1 student notebook
1 copy of **Activity Sheet 2**, **The Strengths of Different Combinations of Magnets**

For every two students

4 flexible magnets, 25 x 20 x 5 mm (1" x ¾" x ³⁄₁₆"), with a 5-mm (³⁄₁₆-inch) hole in the center
2 plastic cups
1 tongue depressor, or wide craft stick
1 jumbo paper clip (paper clip hook)

22 / Measuring Magnets

LESSON 4

25 washers, USS standard No. 10
Cellophane tape (enough to tape together four magnets)

Preparation

1. Prepare the materials for distribution, and duplicate one copy of **Activity Sheet 2** for each student.

   Note: The No. 10 washers can be stacked on a straw to help keep them organized. To avoid counting out washers, use the fact that a stack 12 cm (4¾ inches) high contains 100 washers. Divide the stack in half, then divide each half in half to distribute approximately twenty-five washers to each team.

2. The Student Activity Book contains diagrams and a description of one way to perform this experiment. Instead, you may want to ask the students to find a fair way of their own to test the weakness (or strength) of magnets. Then you could ask several groups to perform the same experiment and compare their results.

Procedure

1. Begin by asking students to imagine that they are the person who is responsible for buying magnets to hold a cabinet or refrigerator door closed. What should they look for in a magnet? How will they know whether a magnet is too strong, too weak, or just right? Discuss the students' ideas of ways to measure how hard a magnet pushes or pulls.

2. Introduce the idea of a controlled experiment or a "fair test." You may want to tell a story such as the one about arm wrestling in the last paragraph of the teacher's **Background**. Help students think of fair ways to decide which of two magnets pulls more gently. It is more important that students air their views than that they reach consensus.

3. Ask students what they think sticking two magnets together will do to the strength of the magnets. Will they cancel each other out or will their strength be combined? Do they think that three magnets together will be stronger than two? Ask how they could find out.

4. First, describe the materials that students will use to test their ideas, then discuss the advantages of working together and sharing their materials and ideas. Then ask the students to help you demonstrate on the chalkboard how to set up a data table for recording results. Tell them to make a similar table in their notebooks. Remind them of the importance of labeling the columns and putting a title on their paper so that they will know what the numbers mean. The example below could be used as a model.

The Strengths of Different Combinations of Magnets

Number of Magnets   Number of Washers Held
                                                   (Magnetic Strength)

5. Distribute all of the materials for the experiment, except **Activity Sheet 2**, to each pair of students. Observe how the students interact with the materials and with each other. Tell them to follow the directions in their

Measuring Magnets / **23**

LESSON 4

Student Activity Book for the "fair test," reminding them to record their observations in the data table they have set up in their notebook.

Final Activities

Distribute **Activity Sheet 2**. Explain to the students how to report their results on a bar graph. Remind them of graphing essentials, such as labeling the axes and writing a title, so that they will know what the graph represents. Help them to set up the graph and get started, then let them finish it as homework.

Extensions

1. Ask the students to predict how many washers they think six magnets will hold. Have them test their prediction and record both their prediction and their actual result.

    Discuss why certain predictions were made. Ask the students to predict how many washers twelve magnets will hold. What assumptions do students make when predicting in this way? (While their predictions might assume that the combined magnets keep getting stronger by the same amount each time that a magnet is added, students will find through experimentation that this is not true. Instead, because each magnet added is farther away from the hook and washers, it will contribute less to the combined strength of the magnets.)

2. Allow the students to repeat the experiment, but with the magnets forced together and held with tape to prevent them from pushing each other apart. Ask them to make predictions of how the outcome of this experiment will compare with their previous experiments. Ask the students to talk about what they think are possible reasons for the results they obtain.

Evaluation

Review the data tables and graphs that the students produce. These items will help you to identify where each student might need help.

24 / Measuring Magnets

LESSON 4

Activity Sheet 2

NAME: _____

DATE: _____

The Strengths of Different Combinations of Magnets

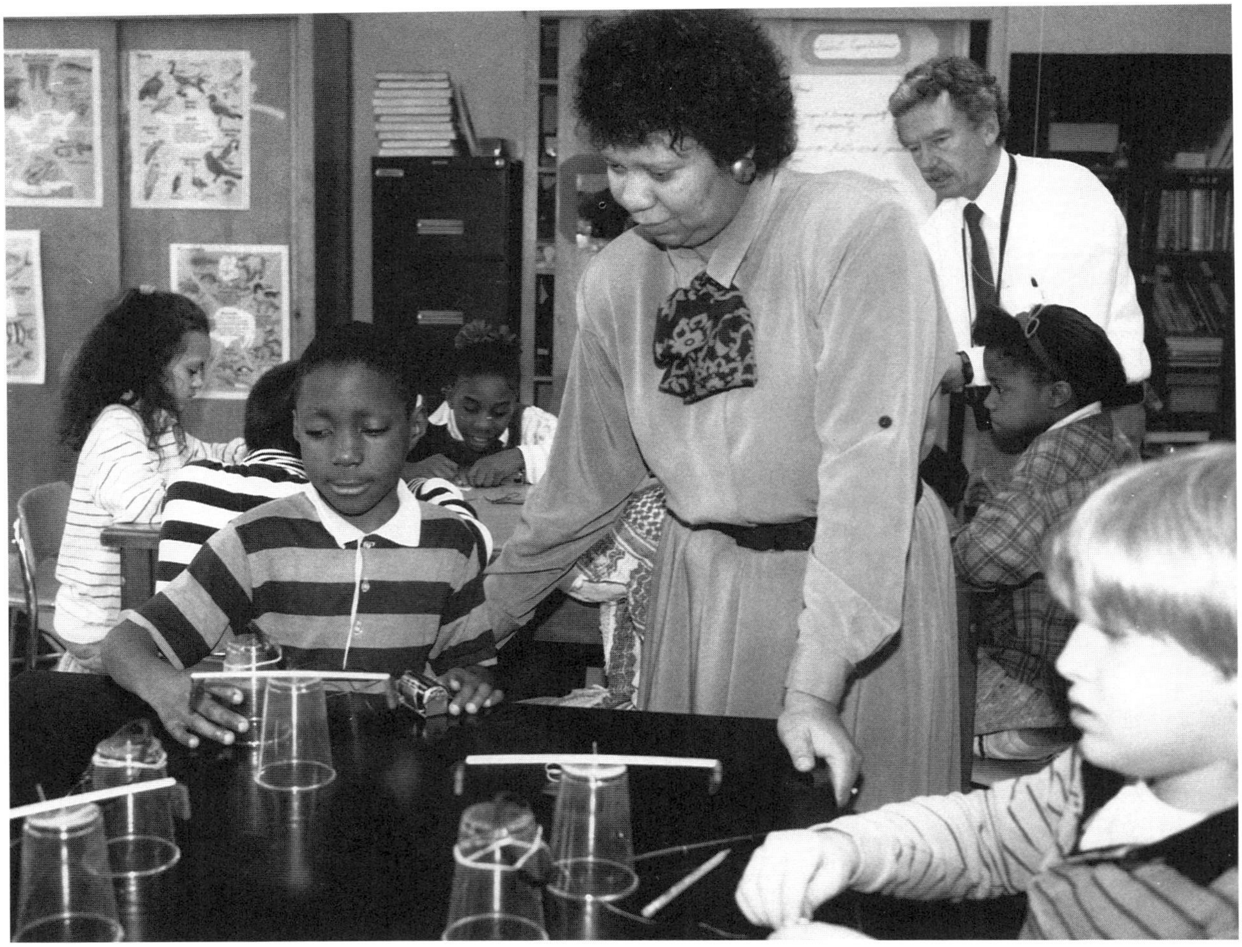

LESSON 5

Building a Compass

Overview

We use magnets in many ways, but at one time, their only use was in compasses to guide explorers in unknown seas and territories. In this lesson, students investigate the behavior of a magnetic compass and learn about its place—and importance—in human history. They then build and adjust their own working compass and begin to explore its characteristics.

Objectives

- Students learn about the historical significance of the magnetic compass and its present uses.
- Students discover the characteristics of a magnetic compass.
- Students develop proficiency in manipulating materials, following plans, and troubleshooting.

Background

Roughly speaking, the earth acts as if it were one big magnet. That is why the magnets in compasses (compass needles) and magnets hanging from strings align themselves the way they do. Compass needles point toward the earth's magnetic poles, which are close to the geographic (rotational) poles.

No one knows for certain what causes the Earth's magnetic field. It is still a mystery. Evidence from old rocks indicates that the Earth's magnetism (magnetic "field") has moved around quite a bit, and even reversed itself repeatedly in the past.

Geologists have been studying the magnetic fields of rocks that formed and cooled at different times and places. They have found that the **polarity** (the direction of a rock's magnetic field) depends on when the rock was formed. One location for observing this phenomenon is the floor of the Atlantic Ocean on either side of the Mid Atlantic Ridge. Here, new rock is emerging very slowly from beneath the earth's crust, cooling, and moving away from the ridge in both directions. The rocks on both sides of the ridge have identical patterns of reversing magnetic polarity as you move away from the ridge.

As you probably know, magnets always have at least two "poles"—two places where the magnetic force is strongest. We label the end of the magnet the "north-seeking pole" if it points to the earth's geographic north pole; we call the other end of the magnet the "south-seeking pole." They could have been named anything, but the names "north pole" and "south pole" have stuck.

LESSON 5

Learning that "opposites attract" often creates a logic problem for students. This means that north and south poles attract. Students may ask, "So how come if the Earth is a magnet, the north pole of a magnet is attracted to the north pole of the Earth?" This is a great question. To be consistent, the only way to explain the problem is to say that the north pole of a magnet points toward the **geographic** north pole of the Earth. By convention, navigators refer to "magnetic north" as the geographic region toward which the north-seeking pole of a magnet points. In fact, however, the magnetic pole now located somewhere in the Arctic Ocean must logically be a "south" **magnetic** pole of the magnet we call Earth, since north-seeking poles are attracted to it. And, the other magnetic pole, located in Antarctica, must logically be a "north" **magnetic** pole.

Figure 5-1

This figure shows the simple but confusing convention that navigators have adopted about the geographic and magnetic poles of the Earth.

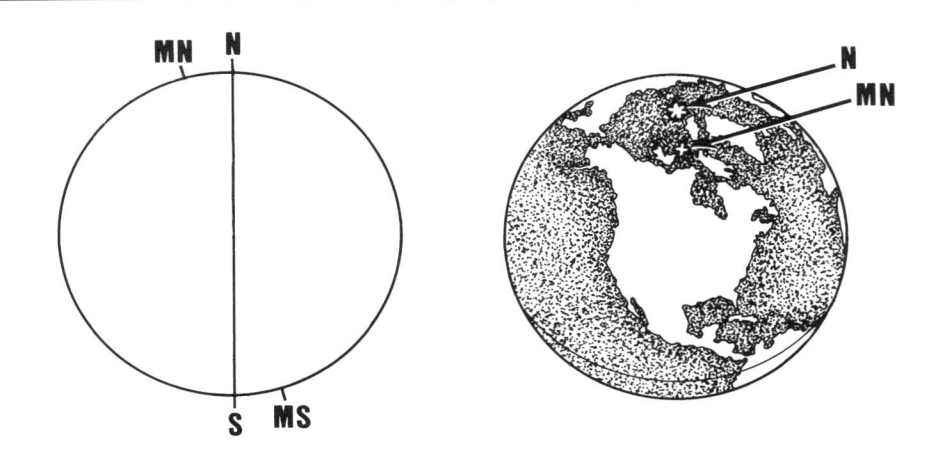

Back to compasses. Because the magnetic poles of the Earth are not located exactly at the geographic poles, a correction is made to a compass reading if a navigator needs to be precise about the direction. And, of course, near the poles of the Earth, a compass is of little value for finding direction.

The use of the magnetic compass enabled ships to sail out of sight of land. With the compass, explorers could travel to areas of the world previously unknown to them. Trade across great oceans began. Better and better maps were drawn to aid navigation. While sailors could use a clock to navigate from the position of the stars and the sun, they still needed the compass to determine the direction in which they were heading on cloudy days and nights.

Management Tip: The compass needles in most inexpensive, commercially produced compasses are only weakly magnetized. Because of this, it is possible for them to become demagnetized or even magnetized with reverse polarity (the north-seeking end of the needle becomes the south-seeking end) by being in close proximity to stronger magnets. You can check easily to see if this has happened to a compass by laying several compasses out on a flat, nonmagnetic surface and observing which way they point.

A compass needle can be repolarized by rubbing a stronger magnet along its length, but it is probably more useful to discuss with students what has happened. This is a great opportunity for students to realize that it is the magnetic properties of the compass needle that make it point the way it does, not the paint or dye on one end of the "needle."

28 / Building a Compass

Materials

For each student
- 1 student notebook
- 2 flexible magnets, 25 x 20 x 5 mm (1" x ¾" x ³⁄₁₆"), with a 5-mm (³⁄₁₆-inch) hole in the center
- 1 plastic drinking straw
- 2 pieces of #22 coated hook-up wire, 20 cm (8 inches) long
- 1 straight pin, 2.5 cm (1 inch) long
- 1 plastic cup

For every two students
- 1 magnetic compass

Figure 5-2

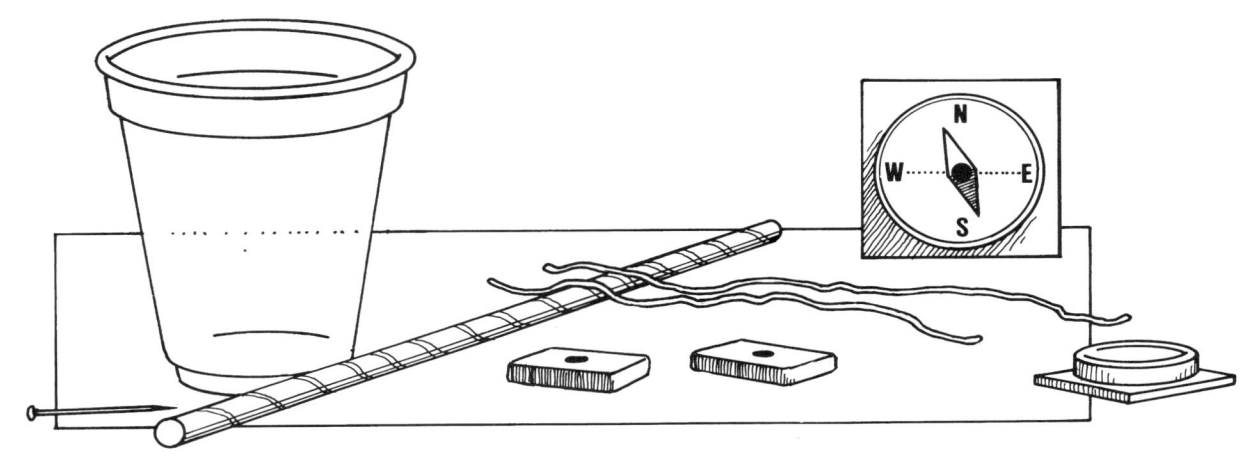

Preparation

1. Prepare the materials for distribution.

2. Prepare a storage area large enough for the students' compasses to be stored without being damaged. One possibility is to remove the compasses from the cups and lay them side by side in a cardboard box.

3. If you have not done so already, you may want to build and balance a straw compass using the student instructions on pgs. 32 through 35 of the Teacher's Guide and pgs. 13 through 16 of the Student Activity Book. Suggestions for troubleshooting are included on the students' instruction sheet.

Procedure

1. Begin this lesson by reviewing with the students the value of the graphs they made. Graphs display information as a picture. Maps are another way to tell someone something without using many words.

   Give the students verbal directions to a nearby landmark—a fire station, beach, or shopping mall, for example. Now, compare the verbal directions you gave with a simple map that you draw on the board. Be sure to include on the map an orienting arrow with an N for north. Which method is a quicker and easier way to communicate directions? Clearly, maps are valuable.

   Ask students what the arrow and N stand for on your map. If they had a paper map, how would they know which way to hold it? How do they

Building a Compass / 29

LESSON 5

know which way is north? Students will respond to this question with a variety of responses. For example, "My street runs north," "Use a compass," "Look at the sun," "I don't know." Use this opportunity to discuss the ways that people navigated before electronics and satellites (by looking at the sun and stars, following ocean currents, following a compass). Ask the students why they think that the magnetic compass was so important.

2. Distribute a magnetic compass to each pair of students. Tell them to hold the compass flat. Have all of the students point in the direction that the compass needles point and comment on the direction in which most students are pointing. Ask them to experiment with the compasses and discuss some of the following questions (listed in the Student Activity Book) with their partners:

- Why do you think most compasses point the same way?
- How can you make the needle point in different directions?
- How could you test to see if the needle (pointer) is a magnet?

After the students have had a few minutes to discuss these questions and manipulate the compasses, collect the compasses and ask specific students what they discussed. Encourage the students to listen to each other. Tell them that they will have time to do some further exploration after they build their own compass.

3. Ask the students to pick up the materials they will need to build a compass. Tell them to follow the instructions for building and balancing a compass on pgs. 13 through 16 of the Student Activity Book (pgs. 32 through 35 of the Teacher's Guide). Some students will have difficulty balancing the compass at first, but with some encouragement they will figure it out. Suggest that students who finish quickly help students who ask for assistance.

Final Activities

1. Ask the students to think and write about how they could make their straw compasses to tell which direction is which. How could they make the compass point in different directions?

2. Ask students to think of other ways they could make a compass. Tell them to write their ideas in their student notebooks.

Extensions

1. Attach a compass to the end of a rolled-up paper tube. Have the students close or cover one eye and look straight down through the tube at the compass. The compass should be about the only thing that the student is able to see.

Have the students follow a partner's directions to navigate from one place to another in the room. Write the directions on a 3" x 5" card. The following directions are a sample:

Begin at your desk. Take 3 steps north. Now, take 1 step east. Next, take 8 steps south, then 3 steps west. Take 5 steps north and 2 steps east. Where are you? (Back where you started.)

This activity could be coordinated with a mathematics lesson on angles, shapes, or problem solving. Ask the students to make up directions with a certain number of steps to get from place to place in the classroom. Then, they could draw a map of the directions.

30 / Building a Compass

2. On a map of the world, trace the routes of various explorers and discuss the role that magnetic compasses played in their expeditions. Ask the students to think about the problems that explorers face when navigating in the Arctic and Antarctic regions. (Compasses are useless there and in the summer, when most expeditions take place, the sun appears to go around and around near the horizon, never rising or setting.)

LESSON 5

Student Instructions for Building a Compass

1. Bend the two wires in half around your finger. You are doing this so that the wire will fit snugly inside the straw when you get to Step 5.

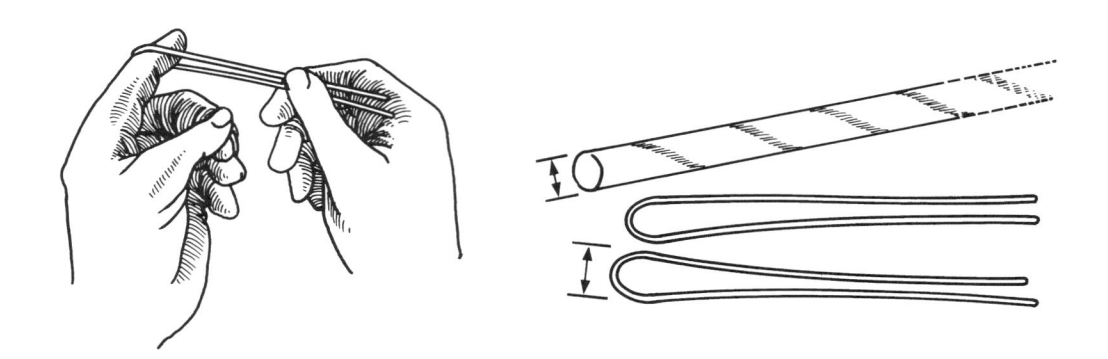

2. Arrange the two magnets so that they stick together side by side, as shown. Push the open ends of one wire one third of the way through the hole in the center of one of the magnets. Now do the same to the other magnet with the other wire.

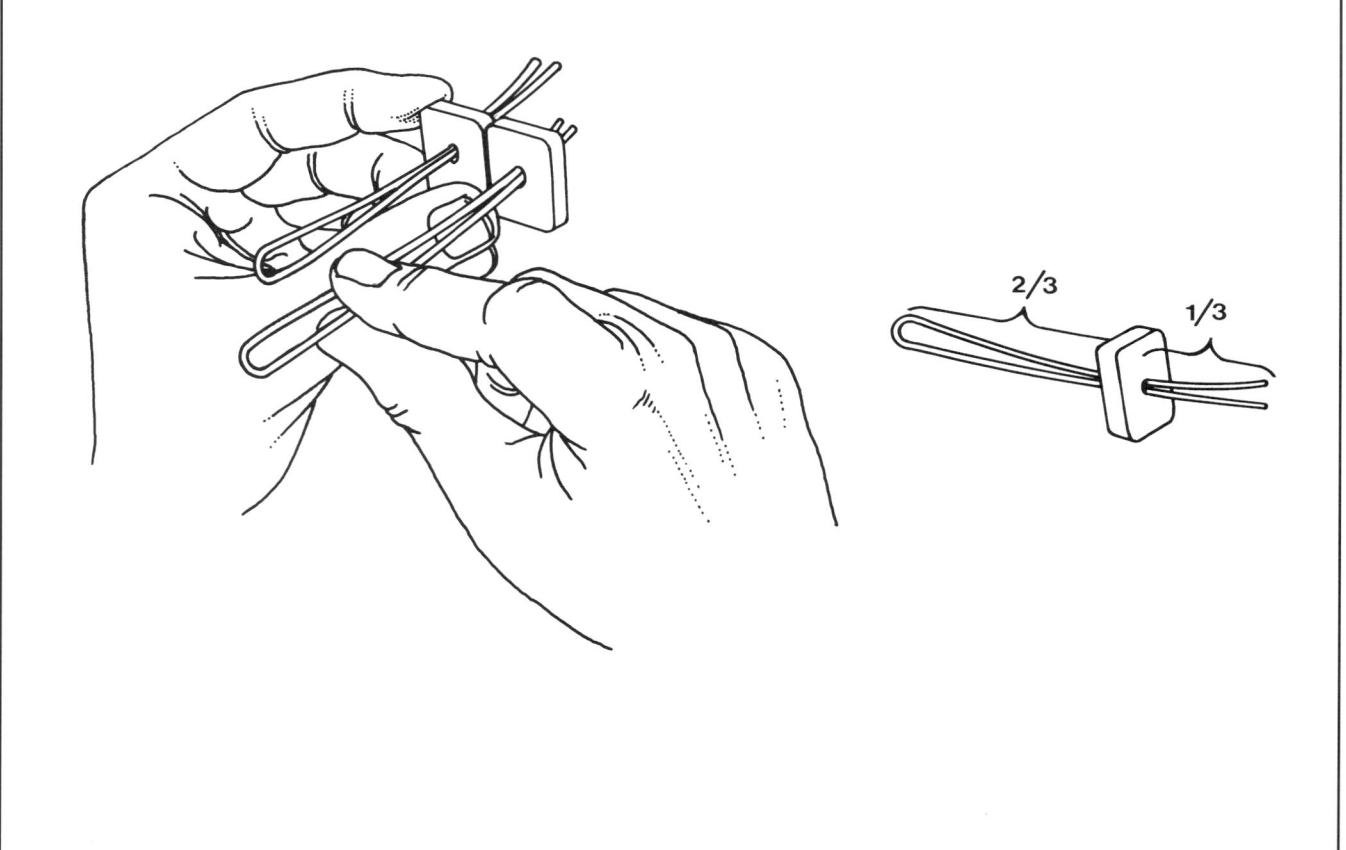

32 / Building a Compass

3. Without removing the wire, separate the two magnets. Pull apart the open ends of the wire and wrap each end around the long sides of the magnet. Bend the looped end up over the top and let it project forward. Now do the same to the other magnet.

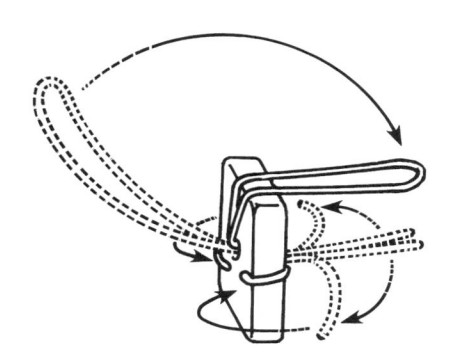

4. Check to see if the sides of the magnets away from the loop will attract each other. If they repel, go back to Step 2, remove the wire from one of the magnets, turn it over, and rewire it.

   The magnets should stick together, with the looped ends pointing in opposite directions.

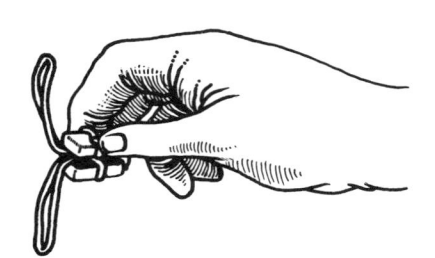

5. Insert the looped ends into the ends of the straw.

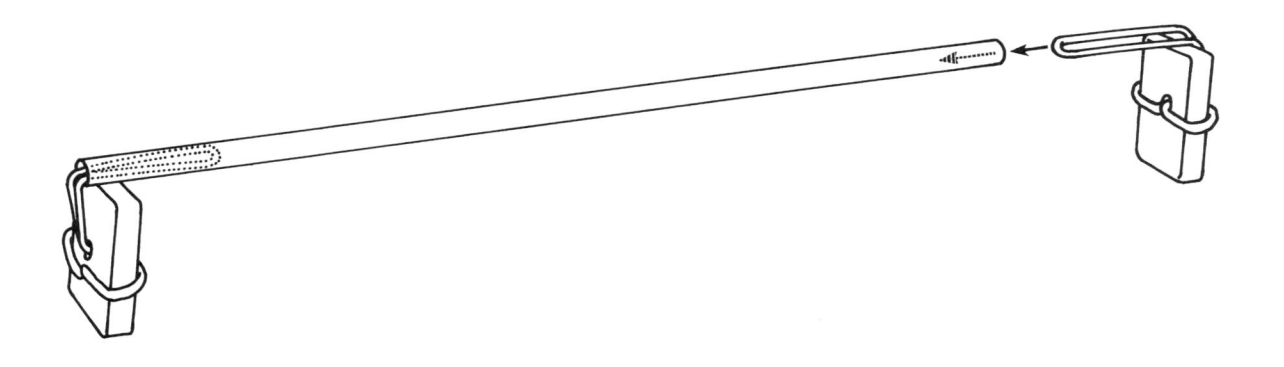

Building a Compass / 33

LESSON 5

6. Balance the straw on your finger and push the pin down through the straw at the place where it balances. The tip of the pin should project about 5 mm below the bottom of the straw. Do not make more than one or two holes in the straw or the pin will slip.

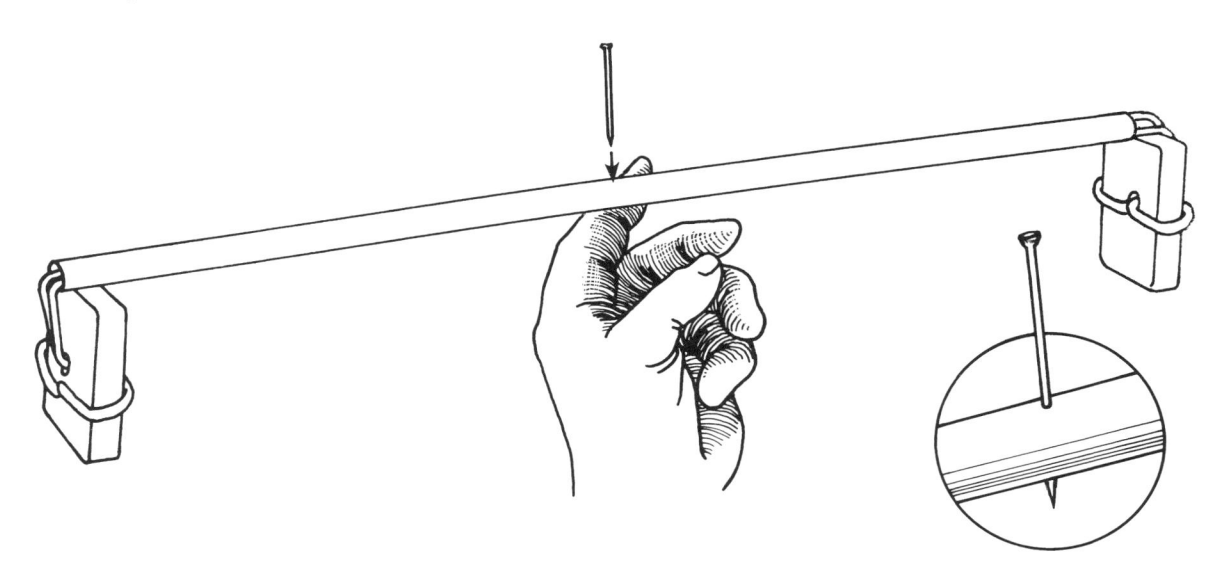

NOTE: Do not push the pin into the plastic cup. Use the tips below to balance your compass.

7. Balance the straw across the top of the inverted plastic cup. Adjust the balance of the straw compass by pulling out or pushing in the magnets at the ends of the straw a little bit, as needed.

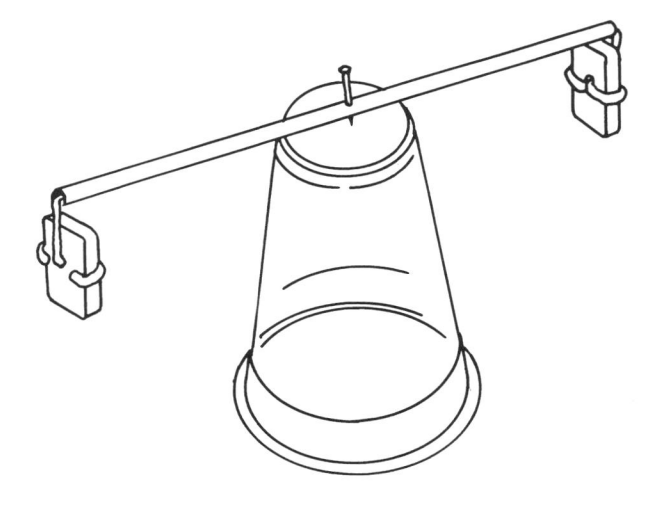

34 / Building a Compass

To balance the compass side to side, rotate or twist one or both magnets slightly.

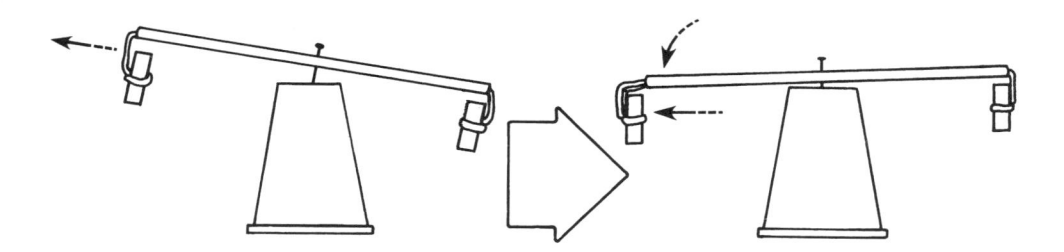

This kind of adjusting is called **troubleshooting**. Be patient and adjust one thing at a time. You will get your compass to balance.

If you have difficulty, keep trying. You will discover how to make your compass work.

Do not disassemble your compass. You will use it again in later lessons.

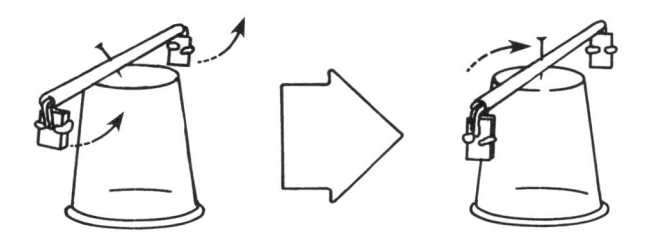

Building a Compass / 35

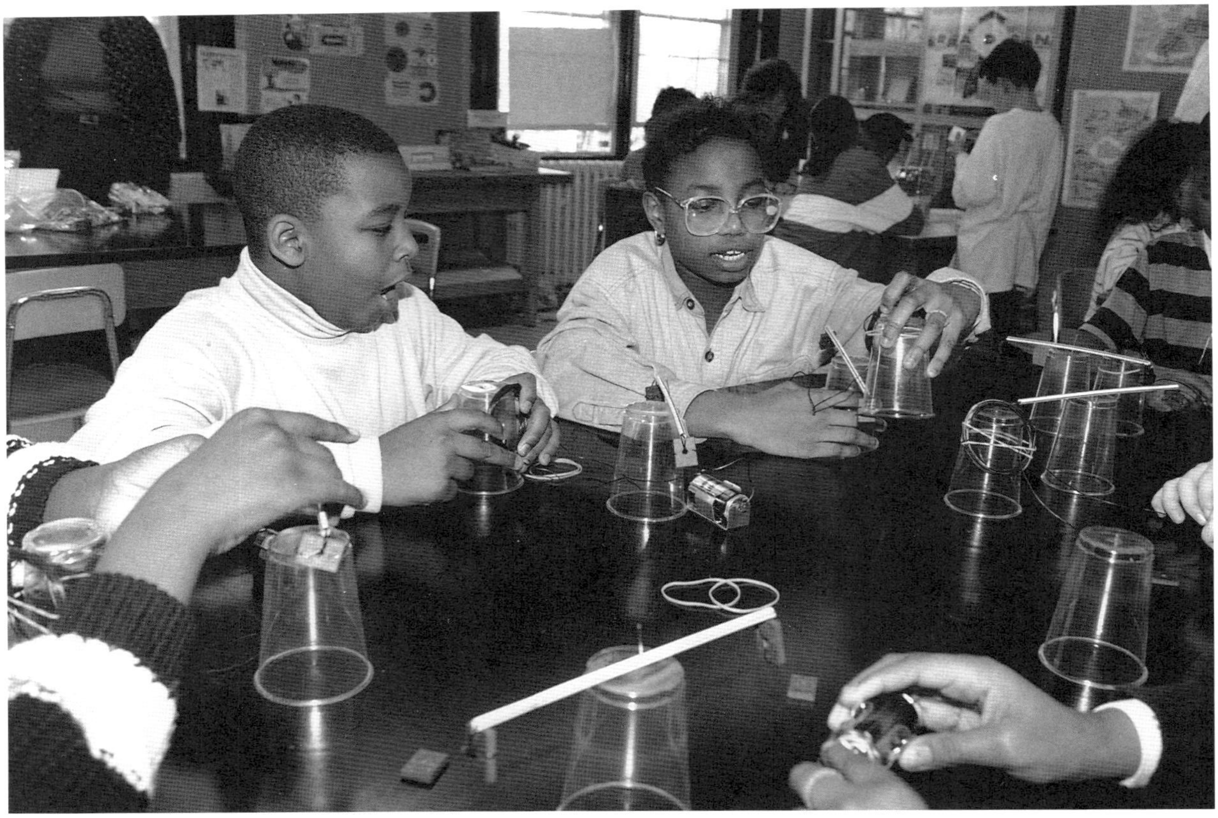

LESSON 6

Using a Compass: Which Way Is Which?

Overview

In this lesson, students experiment with the compasses they built in Lesson 5. They determine which poles of the magnets point north, explore the geographic usefulness of a compass, and make and test predictions of magnetic behavior. The students then use what they have learned about attraction and repulsion to make the compass spin like a motor. This activity is repeated in Lesson 12 using an electromagnetic coil.

Objectives

- Students learn more about the geographic usefulness of a compass.
- Students investigate magnetic poles.
- Students begin thinking about the ways in which magnets can be used to cause motion, as in a motor.

Background

Because the compasses that the students built have two magnets, they also have two north and two south poles. This fact does not interfere with the operation of the compasses, but it may provoke some interesting questions and observations from students as they experiment with them.

Management Tip: If a compass does not point north and south, this probably is because a magnetic object is nearby or because one of the compass magnets was installed incorrectly. (See Step 4 of the Instruction Sheet on pg. 14 of the Student Activity Book; pg. 33 of the Teacher's Guide.)

Materials

For each student
- 1 student notebook
- 1 straw compass (from Lesson 5)
- 2 colored stickers, the same color
- 1 flexible magnet, 25 x 20 x 5 mm (1" x ¾" x ³⁄₁₆"), with a 5-mm (³⁄₁₆-inch) hole in the center

For the class
- 1 large map or globe of the world

LESSON 6

Preparation

1. Display the large map or globe.
2. If you have not already done so, you may want to take a little time to locate the north side of your classroom and to investigate the behavior of a straw compass.

Procedure

1. Begin the lesson by allowing the students to finish assembling and balancing their magnetic compasses.
2. Demonstrate the way that you would like the students to test their compasses by releasing a compass from several different positions. Simply point it in any direction and let go. The compass will slowly swing to align north-south. It will be helpful for the students to do this while you are demonstrating.
3. Ask the students to notice the similar alignment of all the compasses. Review briefly the discussion that took place during Lesson 5 about the possible reasons for this strange alignment.
4. Stand on the north side of the classroom. Ask the students to mark the north pole of both magnets with the colorful stickers. Doing this as a class will ensure that everyone does it the same way. It also will allow you to check quickly those students who may be having difficulty.
5. Use the map of the world to discuss world geography with the students and to help them understand how a compass works. Ask questions. For example, "If you were in Australia, which continent would the south pole of a compass magnet point toward?" "Which way would you head to get from South America to Africa?" "Which state would the north pole of your compass point toward if you were in Pennsylvania?"
6. Next, ask the students to work with a partner and his or her compass. Tell them to follow the steps outlined on pg. 18 of the Student Activity Book. Remind them to record what they find in their notebook.

    The steps the students are following will help them to:

    - Notice the interaction between opposite and like poles of magnets when they move the two compasses close together
    - Find and mark the north pole of an unmarked, extra magnet and then test what happens to one of the compasses when the extra magnet is brought near it
    - Use the extra magnet to make the compass spin (this motor-like spinning is repeated in Lesson 12 using electromagnets)

7. Have the students return the intact compasses to the storage area. The compasses will be used again in Lesson 12.

Final Activities

Ask the students to discuss with their partners the method they used for finding the north (or south) pole of the unmarked magnet and the method they used for making the compass turn like a motor. Remind students to write what they found in their notebooks.

Extensions

1. Ask the students to try to design and build another type of compass. Challenge them to think about how they could do this with only one magnet. Tell them to draw a picture of their idea in their notebook and to explain how it works.

2. If, with help, students find a magnet they could use for their new compass, encourage them to build it, then to test it using the same methods they used to test their straw compasses.

3. Ask the students how they think the compass would behave if the length of the straw were changed. Allow students to experiment in small groups and arrive at their own conclusions about the effect of the length of the straw on the behavior of a compass.

   This type of open-ended investigation also could be used to examine the effect on a compass of turning the magnets sideways, edgeways, or all the way around.

4. Discuss the built-in "compasses" of certain animals. There is evidence that a number of animals other than human beings, among them certain bacteria, some turtles, and pigeons, use the earth's magnetic field to navigate. See the **Reading Selection** on pg. 40 (pg. 19 in the Student Activity Book).

Evaluation

Compare what the students write and think about magnets at this point with what they wrote and thought in Lesson 1. Look for students who do not know that the direction in which the compass points is determined by magnetism. These students missed something and probably will need repeated exercises with the materials in Lessons 2 through 5. Students who are still unsure about the meaning of "opposites attract, likes repel" also will need additional coaching. An understanding of this concept is necessary in order for the students to make sense of their experiences with electromagnetism and motors in later lessons.

LESSON 6

Reading Selection

Do Animals Use Magnetism?

If you were lost in the middle of the woods and couldn't see the sun, you might use a compass to try to decide which direction to take. A magnetic compass needle lines itself up with the Earth's magnetic field and points roughly north and south; from that, you can figure out east and west, too. Because this works fairly well, people have been using magnetic compasses to find their way for about 1,000 years.

But how do animals find their way? How do they navigate when it is cloudy? You probably know that many animals rely on their sense of smell to keep track of where they have been and where other animals are. However, some animals migrate (travel from one place to another), regularly covering hundreds or even thousands of miles each year. It seems unlikely that animals could repeat such long trips accurately if they were relying only on their sense of smell, so scientists have been looking for evidence of what else animals may use to navigate. There are scientific investigations of whether animals use the sun and moon, the earth's magnetic field, and recognition of landmarks to repeat their long journeys.

No one can be sure yet whether animals actually use the earth's magnetic field to help them find their way. Some experiments seem to indicate that they do, and some that they do not. It is a question a number of people are trying to find answers to. How, for example, do homing pigeons find their way home—even on cloudy days? Some studies have found out that it may have something to do with magnetism because pigeons get confused when they fly through areas with particularly strong and irregular magnetic fields. Other studies have indicated that pigeons may use the sun and their sense of smell to navigate. Possibly, they use several different things together or at different times.

If animals like homing pigeons do use magnetism in some way, how do they do it? Obviously they do not have compasses like people do. Researchers are still trying to think of experiments that will reveal how the pigeons do it. And how sea turtles make their way. Sea turtles have shown that, to navigate, they will swim into waves to help them know they are headed away from a beach. They also seem to swim toward bright lights at night. But how do they navigate on dark nights when they are not near a beach? How do they know which way to go? Experiments are underway to find out if sea turtles use magnetism to navigate.

Although people who do scientific research into possible connections between magnetism and animal navigations are still puzzled about pigeons and sea turtles, they are more certain that bacteria are able to sense magnetism and that they may actually use it to find their way. Several experiments have shown that bacteria respond to magnetic fields. And when they swim through water near a strong magnet, they all swim in one direction—depending on where the magnet is placed. These are some of the reasons people continue to wonder about whether larger animals also can detect and use magnetic fields.

Using a Compass: Which Way Is Which? / 41

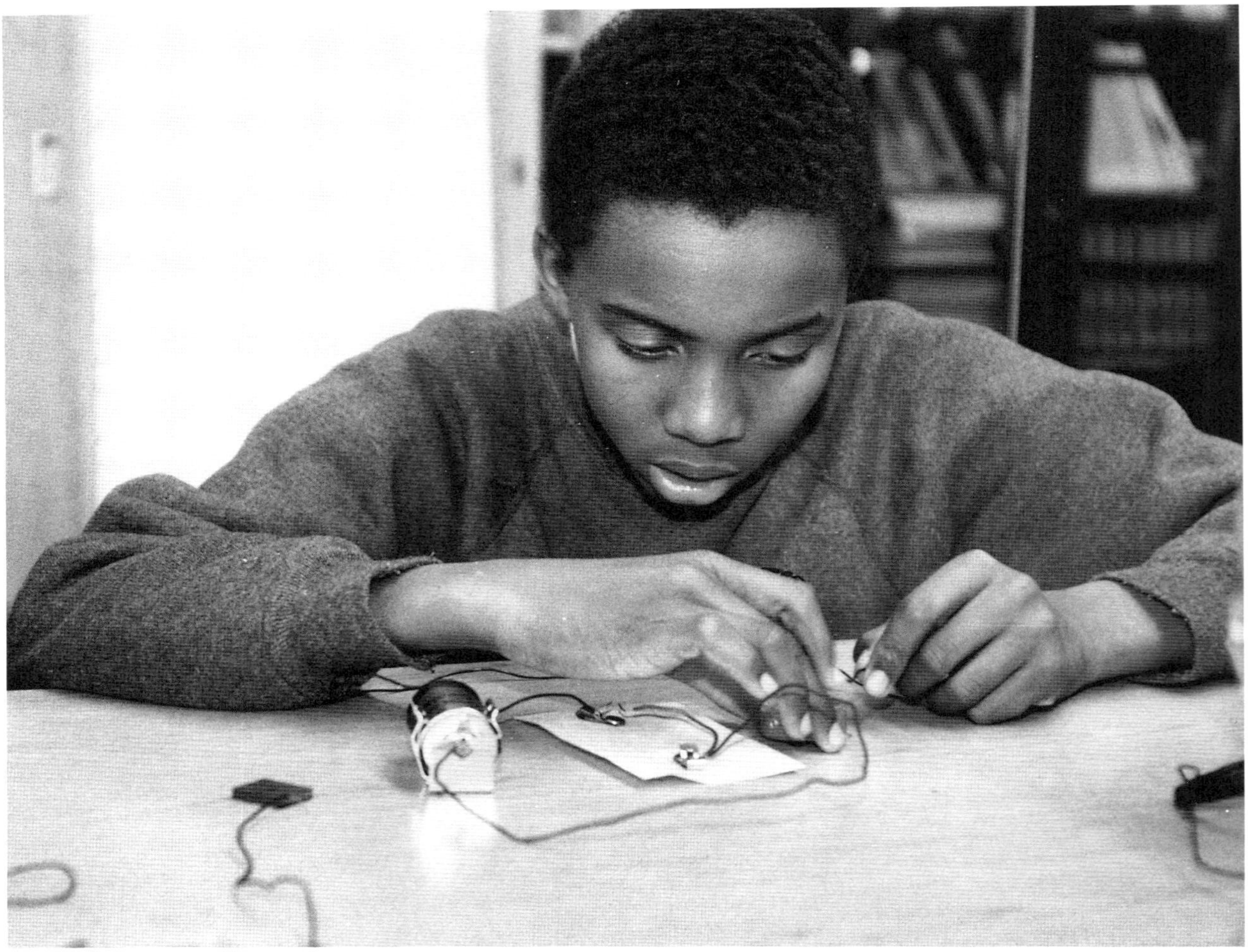

LESSON 7

Creating Magnetism through Electricity

Overview

In this lesson, students discover another way to create the magnetism they have been investigating in Lessons 2 through 6—through electricity. They build an electric circuit, which lights a bulb and causes a compass needle to move. The teacher has an opportunity to find out what they think about electricity. Please be sure to make students aware of safety rules for working with electricity. See the **Safety Reminder** on pg. 44.

Objectives

- Students express the ideas they have about electricity.
- Students observe several characteristics of an electric current by experimenting with an electric circuit.
- Students learn by experimentation that an electric current causes magnetism.

Background

People have long suspected that magnetism and electricity are related. The fact that static electricity produces an attractive and repulsive force like magnetism suggested a link between the two. However, not until 1819 did anyone actually demonstrate a connection. The connection was made with current electricity, which had come into use in 1800 with the invention of the electric battery.

Hans Christian Oersted, a Danish science teacher, noticed that an electric current from a battery had an effect on a compass needle nearby. He did not expect this; it happened while he was attempting to demonstrate to his class that the electric current would **not** affect the compass. Shortly thereafter, electromagnets, motors, generators, and many other devices were produced. (You may wish to refer here to the timeline at the end of **Appendix G**.)

Management Tip: In this and ensuing lessons, students may discover that wires get warm if a circuit is connected for more than several seconds. They also may discover that it is possible to generate higher temperatures by using a thin wire or more batteries. You may want to explain that this is how a light bulb works—electric current is pushed through a tiny tungsten filament to produce heat and light. However, if the students do extended experiments with heat generation, the lifetime of the batteries will be quite short. (Please see **Appendix D** for further information about electric circuits.)

Creating Magnetism through Electricity / **43**

LESSON 7

> **Safety Reminder**
>
> The batteries and circuits that students will investigate in this lesson are not potentially lethal. The voltage provided by even as many as 20 batteries (30 volts) can produce enough current to produce very hot wires, but not life-threatening shocks. Household electricity is potentially lethal, however, and students should be instructed explicitly not to experiment with it. The Student Activity Book (pg. 22) notes to students that they should: 1) never experiment with electricity from wall plugs because it is dangerous and potentially lethal; 2) never work with appliances or lights that are plugged into the wall. Review of these safety notes takes place in Step 1 of the **Procedure** for this lesson.

Materials

For each student
- 1 student notebook
- 1 battery and battery holder
- 1 bulb and bulb socket
- 1 switch made of 1 bulb socket and 1 bolt, 12 mm x 8 mm (½" x ⅜")
- 3 pieces of #22 coated hook-up wire, each 20 cm (8 inches) long
- 1 piece of #22 coated hook-up wire, 80 cm (32 inches) long
- 1 magnetic compass

For the class
- 1 large sheet of newsprint or other paper
- 1 wire stripper tool

Figure 7-1

Preparation

1. Cut three 20-cm (8-inch) pieces and one 80-cm (32-inch) piece of wire for each student. Include a few extras for unforeseen circumstances.

2. Strip approximately 1 cm (⅜ inch) of insulation from each end of the wires using the wire-stripper tool. You may want to enlist students to help with this task. (See **Appendix E** for information on how to use a wire-stripper tool.)

3. Assemble the materials for distribution.

44 / Creating Magnetism through Electricity

LESSON 7

Procedure

1. Begin the lesson by asking the students what electricity is used for. List their ideas on a large sheet of newsprint or other paper so that you can assess later how their thinking has changed.

   Explain to the students that they will be working with electricity in order to learn more about it. Review with them the **Safety Reminders** on pg. 22 of the Student Activity Book. These are summarized at the end of the **Background** section for this lesson.

2. Distribute the materials and ask the students to use them while following the instructions on pgs. 23 and 24 of the Student Activity Book (pgs. 46 and 47 of the Teacher's Guide).

   The instructions lead the students through the following activities:

   - Building a circuit that will light a bulb
   - Incorporating a switch into their circuit to control the electricity that lights the bulb
   - Building a **"short circuit"** by replacing the bulb with a wire and investigating the effect on a magnetic compass of electricity flowing through the wire
   - Using the magnetic compass to detect whether current is flowing through the other wires of the circuit

3. Observe the students as they work with the materials. Encourage them to find out things for themselves and to think about what they are doing and how they think electricity and magnetism work. Resist the temptation to help the students by doing things for them.

4. Tell the students to return their materials to the storage area.

Final Activities

1. Ask the students to write in their notebooks what they discovered while working with their circuits and compasses. Ask them to draw pictures of the circuits they built and to use arrows to indicate where electric current is flowing.

2. Encourage the students to discuss their findings with each other. If different outcomes become apparent, suggest that they design an experiment to try to resolve them.

Extension

Challenge the students to predict and then test whether and where current is flowing through various circuit arrangements. Allow them to design their own arrangements.

Creating Magnetism through Electricity / **45**

LESSON 7

Student Instructions for Working with Electricity and Magnetism

1. Using two of the 20-cm wires, connect the battery and battery holder with the bulb and bulb holder so that the bulb lights up. This type of connection is called an electric circuit.

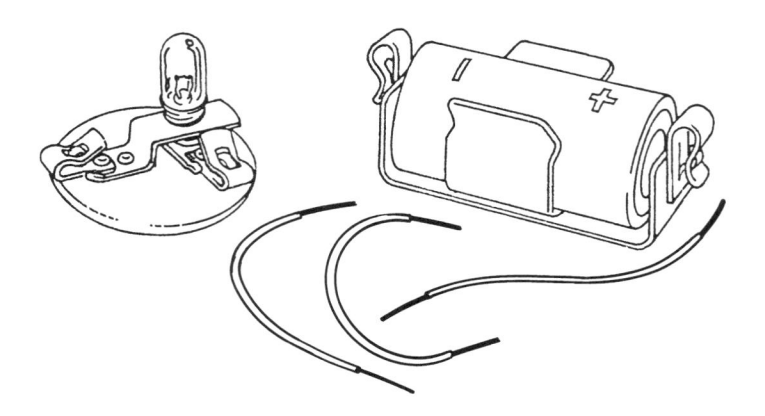

2. Next, try to connect the circuit so that the switch is included. You will be able to use the switch to turn the light on and off.

3. Think about where you believe the electricity is flowing when the switch is on. What happens when the switch is off?

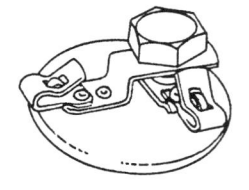

4. Now, take the bulb and the bulb holder out of the circuit and replace them with the 80-cm wire.

    Place the compass under the wire so that the wire runs in the same direction as the needle is pointing. You may have to bend the wire to do this.

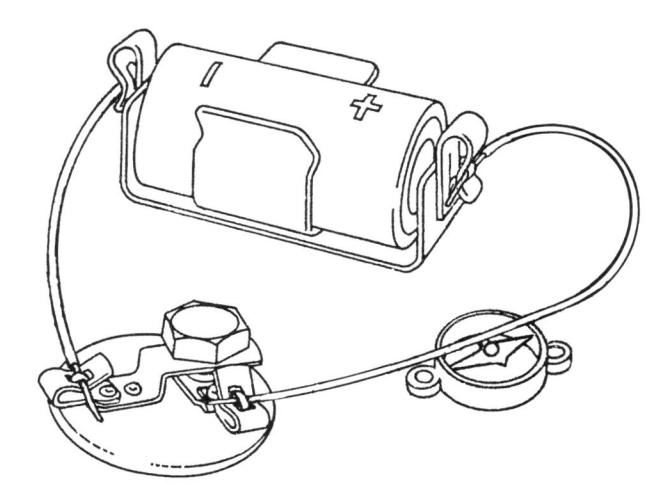

5. Wait a moment for the compass needle to settle down, then turn the switch on carefully. What happens? Why do you think this happens?

    What happens when you turn the switch off?

6. Try testing the other wires in your circuit by placing the compass under them. What happens when you turn the switch on and off?

    Predict which wires have electric current flowing through them. How can you tell?

Creating Magnetism through Electricity / 47

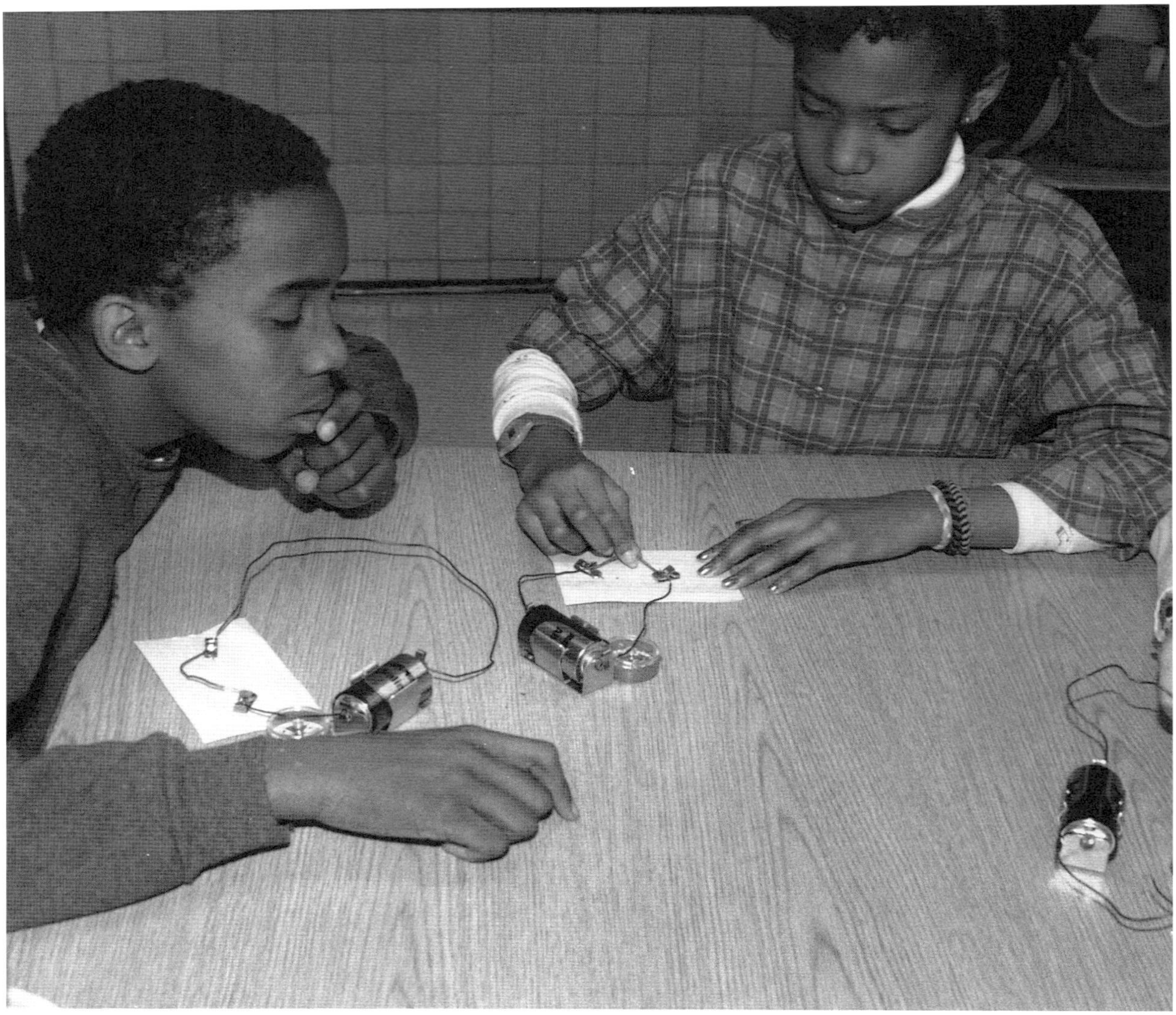

LESSON 8

Making Magnets with Electricity

Overview

Making the connection between electricity and magnetism can be difficult, especially when it is discussed in abstract terms only. In this lesson, students build on their experiences from Lesson 7 to manipulate a magnetic coil of wire and build an electromagnet. The electromagnet is the subject of controlled experiments designed in Lesson 9, conducted in Lesson 10, then reported to the class in Lesson 11. The activities with motors in later lessons also use the electromagnet—a necessary part of an electric motor.

Objectives

- Students learn that a coil of wire with electricity flowing through it has magnetic poles.
- Students build an electromagnet and begin to formulate their own questions about how it works.

Background

Electromagnets are made by coiling wire around an iron core and then passing electricity through the wire or by just making a coil of wire and then passing electricity through the coil. Since each loop of wire becomes magnetic when current flows through it, coiling makes the electromagnet stronger; each loop acts as a little magnet and contributes a bit to the total magnetic force. In this way, extremely powerful magnets can be produced.

Figure 8-1

An electromagnet

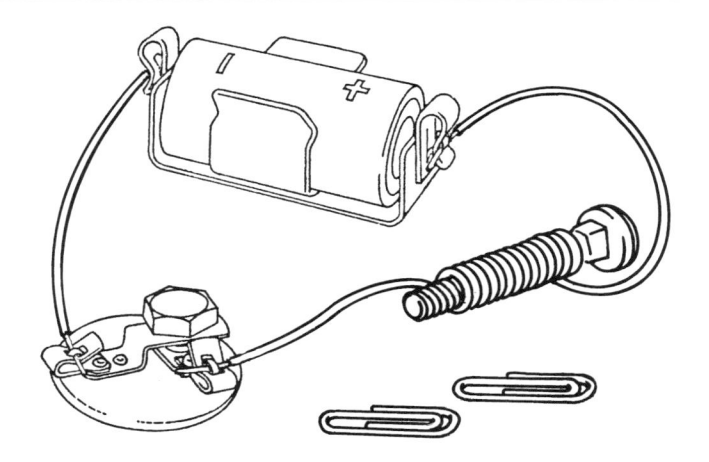

LESSON 8

Iron is the material most commonly used as a core because of its special characteristics. Sometimes, the iron core remains slightly magnetized even after the electricity stops flowing through the wire coiled around it. This is called residual magnetism, and it can create problems for the students' understanding of electromagnets. See the **Note** below.

Note: A few students may falsely interpret residual magnetism as being caused by the wrapping of the wire around the core instead of by the electricity flowing through the wire. If this becomes an issue in your class, challenge students to figure out the source of the residual magnetism or help students experiment to find its source. They can do this by dropping two cores on the floor (this will remove any residual magnetism), then comparing the effect—on a paper clip—of a core that is only wrapped with wire to the effect on a paper clip of a core that is wrapped with wire *and* connected to a battery.

Materials

For each student
- 1 student notebook
- 1 battery and battery holder
- 1 bulb and bulb socket
- 1 magnetic compass
- 1 piece of #22 coated wire, 80 cm (32 inches) long
- 3 pieces of #22 coated hook-up wire, each 20 cm (8 inches) long
- 1 switch
- 1 steel bolt, 8 cm x 6 mm (3" x ¼")
- 3 No. 1 paper clips

Figure 8-2

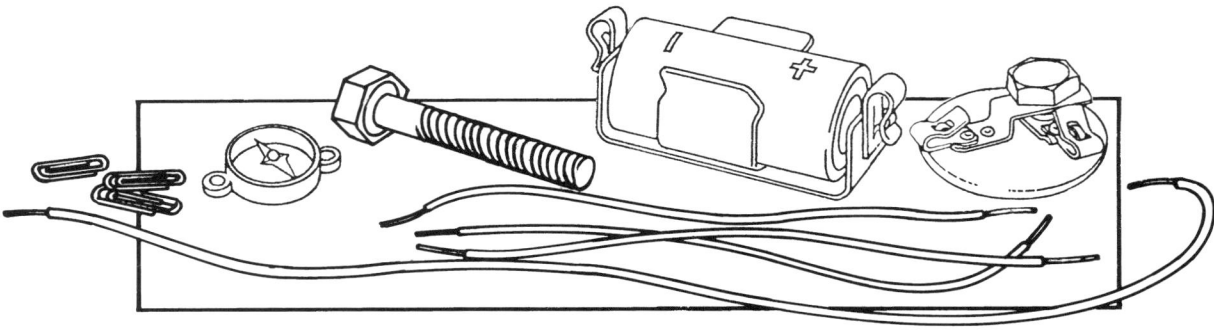

Preparation

The materials for this lesson are the same as those for Lesson 7, except for the addition of a steel bolt and three paper clips.

Procedure

1. Begin by asking the students some open-ended questions; for example, "Where do you think magnets come from?" and "What ways can you think of to make magnets?" You may want to take some time to let the students talk about their ideas and the additional questions that they generate. How can you find out if rubbing something with a magnet makes it a magnet? Does it work for everything or only some things? How do you know that electricity causes magnetism?

50 / Making Magnets with Electricity

2. Explain to the students that they will be exploring ways to use electricity to make magnets. Ask them to look back through their notebooks at what they have written and drawn about magnets and electricity. Pair students briefly and let them talk with each other about what they have learned.

3. Ask the students to use the instructions on pgs. 27 and 30 of their Student Activity Book (pgs. 52 and 55 of the Teacher's Guide) to try and determine how magnets can be made using electricity. Distribute the materials and observe what the students do.

   In this activity, the students are working with the following phenomena:

   - The way a compass responds to a copper wire when electricity is and is not flowing through it (this is a review of Lesson 7)
   - The magnetic poles of a coil of wire, as revealed by a compass
   - How the magnetic poles of a coil of wire reverse when the direction of the current being pushed through the coil is changed by reversing the battery
   - Building an electromagnet and using it to pick up and drop paper clips.

4. Ask the students to return their materials to the storage area.

Final Activities

1. Ask the students to draw a picture in their notebooks of how electricity can make magnetism. Have them write about their drawings.

2. Show the students the entry, shown below, that a scientist made in a notebook 150 years ago. Ask the students if they can see any similarities between what "the scientist" did and what they have done. Why do you think the scientist drew pictures in the notebook? What do you think the purpose of writing in a notebook is?

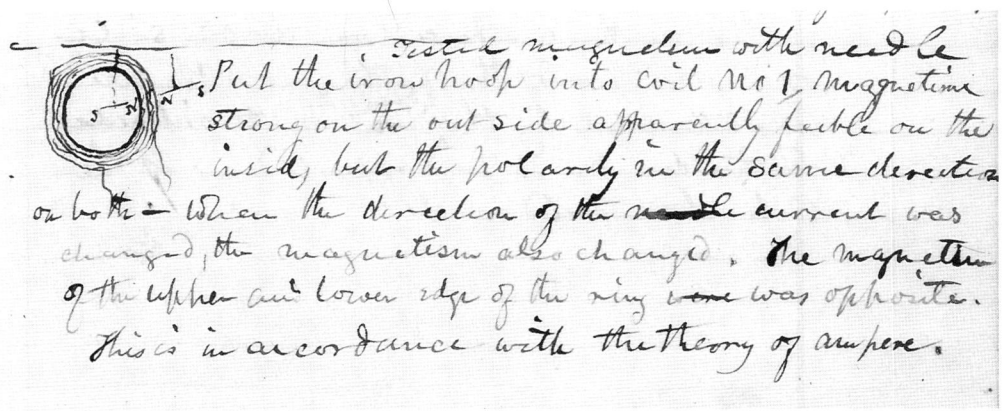

Figure 8-3

Smithsonian Institution Archives; Record Unit 7001; Joseph Henry Papers.

"Tested magnetism with needle. Put the iron hoop into coil number 1. Magnetism strong on the outside, apparently feeble on the inside, but the polarity in the same direction on both. When the direction of the current was changed, the magnetism also changed. The magnetism of the upper and lower edge of the ring was opposite. This is in accordance with the theory of Ampere."

Making Magnets with Electricity / 51

LESSON 8

Student Instructions for Making Magnets with Electricity

1. Begin by connecting a circuit the way you did in **Lesson 7**.

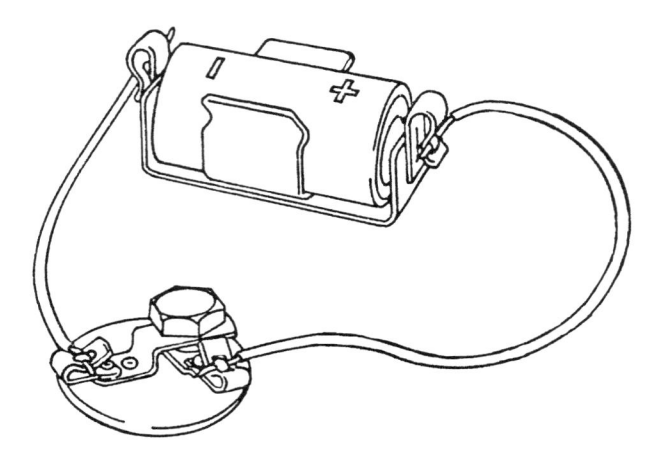

2. Test the wire with the compass to see if it is magnetic. What do you remember about what happens to the compass when you turn the switch on and electricity flows through the wire?

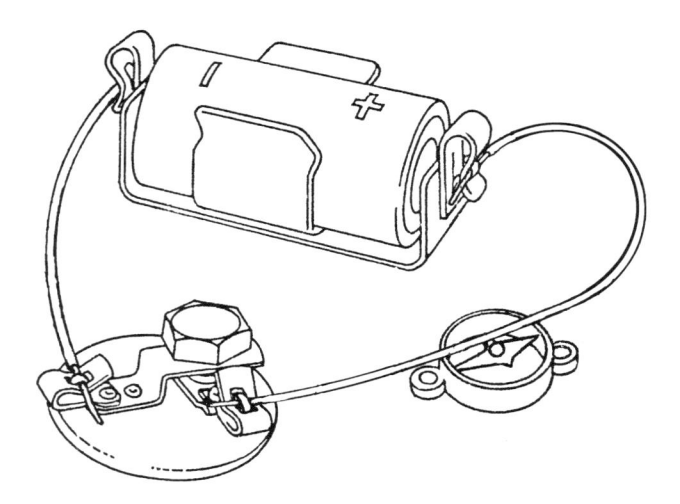

3. Make the long wire into a round **coil** by wrapping it around your fingers. Slip the coil off your fingers. Then twist the ends of the wire around the coil so that it does not unwind. Make sure you have enough wire at each end (10–15 cm, 4–6 inches) so that you can connect the coil into your circuit.

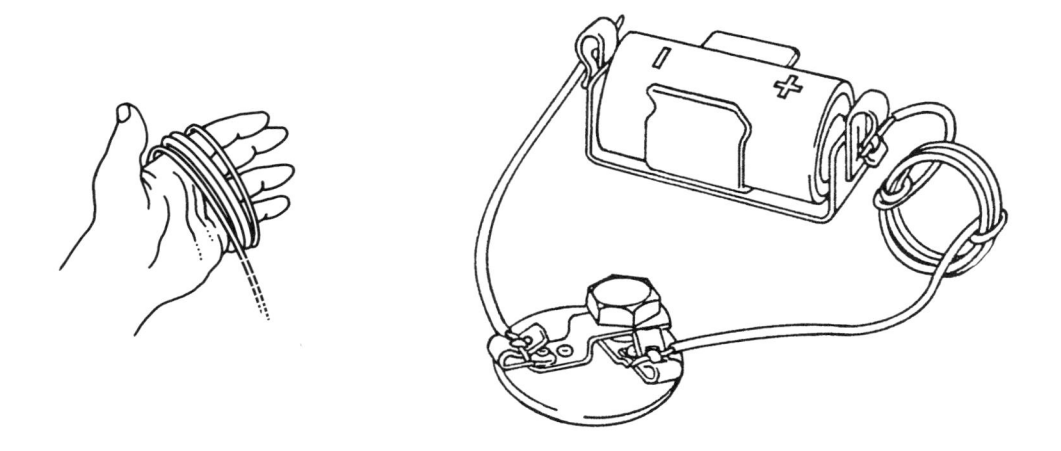

4. Notice what happens to the compass when you hold it close to the coil and turn the switch on. How can you tell whether the coil becomes magnetic?

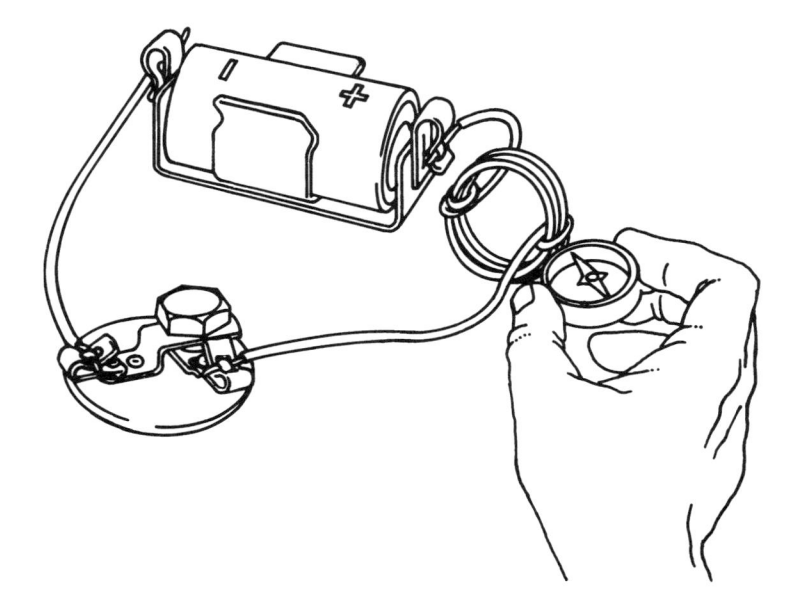

Making Magnets with Electricity / 53

LESSON 8

5. All magnets have two **poles**—places where the magnetic force is strongest. Use the compass to determine where the poles of the coil are when the switch is on and the coil is magnetic.

   ■ Remember that the north pole of the compass magnet is attracted to the south pole of another magnet or electromagnet.

   ■ Draw a picture and label it to describe where you think the north and south poles of the coil are.

   ■ Explain how you know where the poles of the coil are.

6. What do you think will happen to the electric current flowing through the circuit if you take the battery out, flip it over, and replace it in its holder? What do you think will happen to the magnetic poles of the coil?

   Can you think of a way to find out?

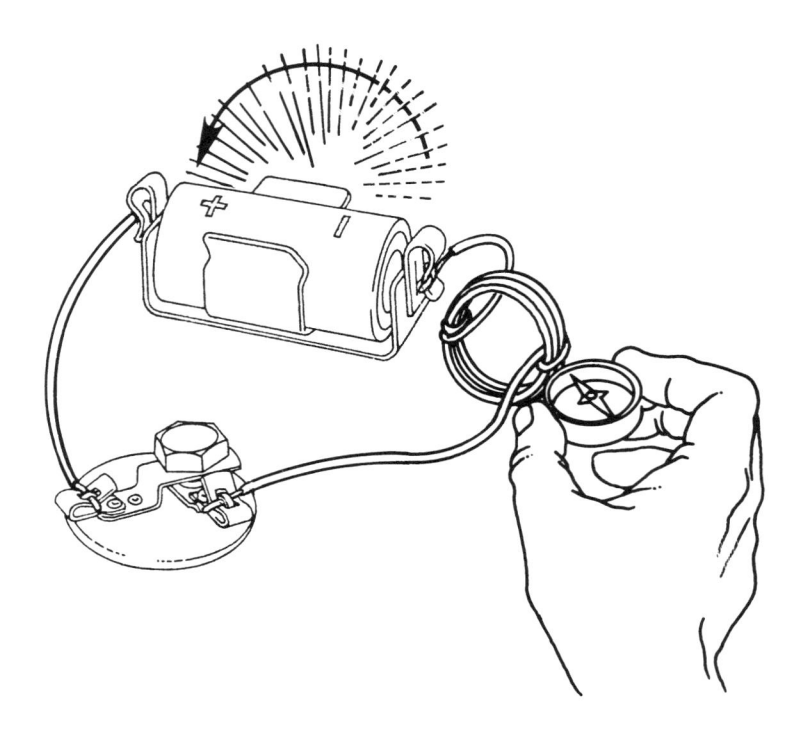

LESSON 8

7. You just built one kind of **electromagnet.** Now try building another kind. Unwind your coil and wrap it around the steel bolt. Leave some wire at each end so that you can connect the wires to the rest of the circuit.

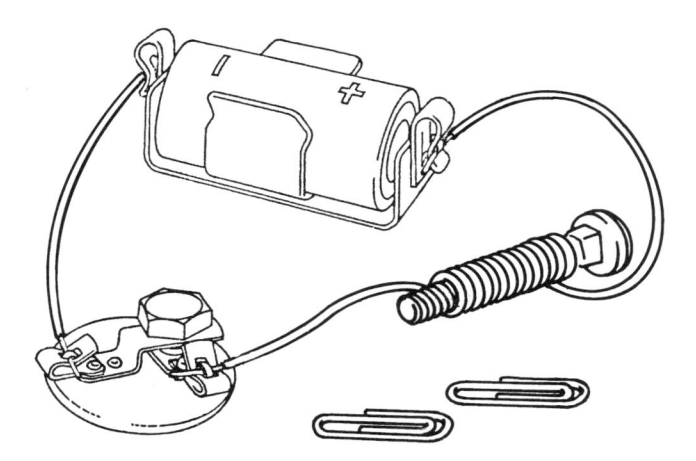

Can you make your electromagnet pick up and drop paper clips by turning the switch on and off?

Note: Remember to turn off the switch when you are not using the electromagnet so that the battery will last.

LESSON 8

Figure 8-4

An electromagnet built in the 1800s that could hold 2000 pounds

Smithsonian Institution Archives

56 / Making Magnets with Electricity

LESSON 9

Planning an Experiment to Test the Strength of an Electromagnet

Overview

In Lesson 4, students learned about controlled experiments or "fair tests" and one way to measure magnetic force. In Lesson 8, they learned about electromagnets. Lesson 9 will give them a chance to plan a controlled experiment that will investigate how changes in some variables affect electromagnetic force. Students work in teams to design the experiment they will conduct in Lesson 10.

Objectives

- Students develop an awareness of at least one historical figure in the field of electromagnetism—Joseph Henry.

- Students identify several variables that they believe may have an effect on electromagnetic strength.

- Students cooperatively design an experiment to test the effect that changing one variable has on electromagnetic strength.

Background

This is the beginning of a series of lessons in which the students plan, conduct, and report the results of a controlled experiment involving the effect of different variables on the strength of an electromagnet. The activities in these lessons contain many opportunities for spontaneous departures from the written lesson plan. Such departures are often very fruitful for students, so you may want to build some extra time, if available, into these lessons.

Figure 9-1

LESSON 9

In this lesson, teams plan their own experiment. In Lesson 10, they conduct it. Finally, in Lesson 11, all teams report to the class what they have found. It might be advisable to alert students now to the fact that they will be reporting their findings to the class as a whole.

You will likely see evidence of students thinking that there is a "right answer" they are supposed to come up with when conducting these (or any) experiments. Try to encourage them to approach this "fair test" with a different attitude, one of discovery.

The variables that the students will be testing probably will fall into one of the following categories. Some of the intricacies of dealing with each are listed. It is not necessary for students to test all of these variables.

- *The number of turns of wire around the core* (the more turns of wire carrying current, the greater the magnetic strength will be—up to a point). However, increasing the length of the wire will decrease somewhat the flow of electricity.

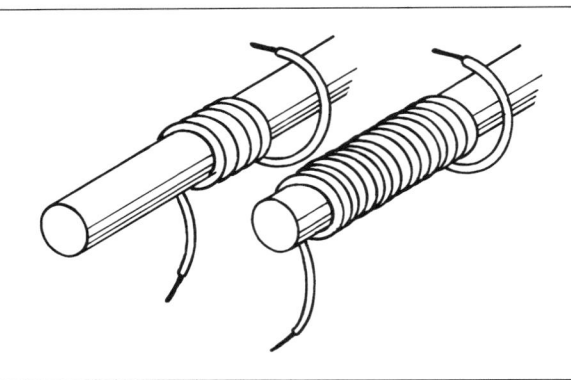

Figure 9-2

The number of turns of wire carrying current has an effect on the strength of the electromagnet

- *The diameter of the core of the electromagnet* (which has an effect on the strength of the electromagnet—students may discover that bigger is not necessarily better). Students may have trouble measuring diameters accurately with a scale. You may wish to use a millimeter scale instead of fractions of inches.

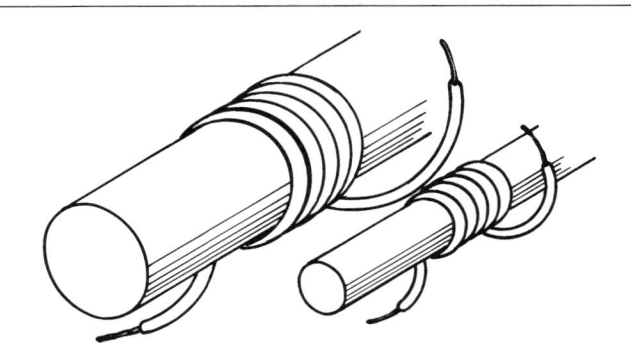

Figure 9-3

The diameter of the core has an effect on the strength of the electromagnet

58 / Planning an Experiment to Test the Strength of an Electromagnet

- *The length of the core of the electromagnet* (which has an effect on the strength of the electromagnet—students may discover that longer is not necessarily better).

Figure 9-4

The length of the core has an effect on the strength of the electromagnet

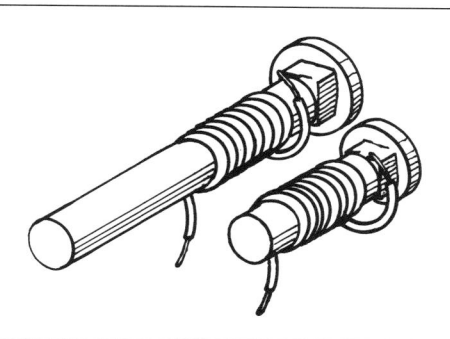

- *The material that the core is made of* (remember that cores containing iron work much better).

Figure 9-5

The material that the core is made of has an effect on the strength of the electromagnet

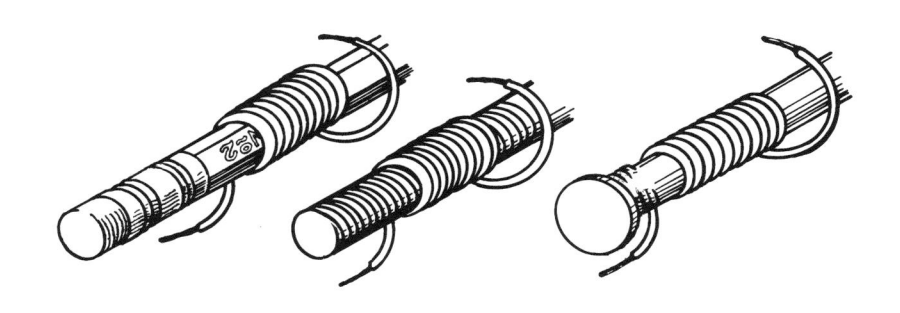

- *The diameter of wire used* (wire thinner than 24 gauge gets very warm when used in a short circuit this way, while wire thicker than 20 gauge is difficult for students to wrap around a bolt and is more expensive; therefore, it is not recommended that this variable be tested by students).

  Note: It is interesting to discuss with students what the "gauge" of wire means. Often, this measure is confusing because the smaller numbers indicate thicker wire. The gauge number (expressed by the symbol "#") represents the number of wires that could be placed side by side in a certain standard distance, like an inch or a centimeter.

Figure 9-6

The diameter of wire used in an electromagnet has an effect on the strength of the electromagnet

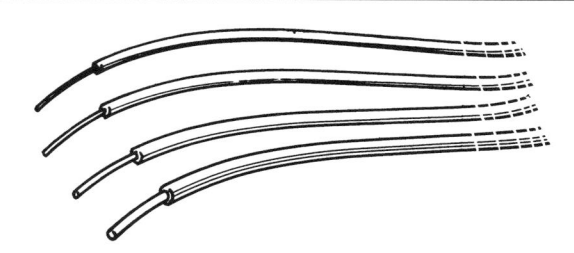

Planning an Experiment to Test the Strength of an Electromagnet / 59

LESSON 9

- *The number of batteries used (in series) to power the electromagnet* (three and four batteries may cause the wire to get very warm if the switch is left on for more than a few seconds; also, the electromagnet may be too strong with four batteries—sometimes, it is hard to get enough washers on the hook to break the grip of the magnet).

Figure 9-7

The number of batteries used has an effect on the strength of the electromagnet

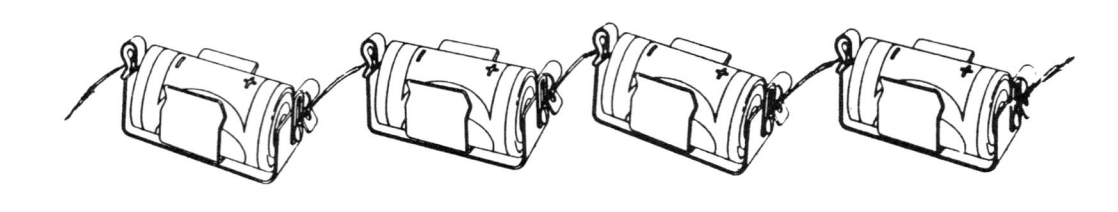

Materials

For each student
- 1 student notebook
- 1 copy of **Activity Sheet 3, Outlining the Team's Experiment**

For every four students
- 1 copy of **Activity Sheet 4, Planning Board**
- 1 envelope containing cards for variables and method (see **Preparation**, Step 2)
- 1 glue stick or bottle of paste

Preparation

1. Reproduce a copy of **Activity Sheet 3** (pg. 64) for each student and a copy of **Activity Sheet 4** (pg. 65) for each team.

2. Cut the colored construction paper into "cards" for each team. For example, for each team, create six yellow cards for the variables, one pink card for the variable to be investigated, and one larger blue card for the description of what the students will do.

   Put the cards for each group in an envelope.

Procedure

1. Begin this lesson by reading to students "Who Is This Story About?" on pg. 63. The story is written so as not to reveal the gender or race of the "mystery person". (It is Joseph Henry.) The idea is for all of the students to feel as if they could be the mystery person!

2. While you read the story, students may interrupt to make guesses about who the person is. The guesses probably will range from Sally Ride to Albert Einstein. Take advantage of each guess to have the students tell each other what they know about the person they suggested. This will take some time, but the history presented by the students is worth the effort.

   If no one knows who you are talking about (and they probably will not), ask the students to see if they can find out who it is by looking in books. At this point, you may want to tell the students that scientists often begin experiments by reading about similar things other people have done.

3. After emphasizing that the person in the story built an electromagnet that could support 2,000 pounds (show students the picture of it on pg. 56), ask them what changes they would make to the electromagnet they built in Lesson 8 in order to make it weaker or stronger. List their ideas on the chalkboard. (For example, "more batteries," "more wraps of wire," "a bigger core.")

4. Explain to the students that they will be working as a member of a four-person team to find out what effect changing one variable has on magnetic strength. You may wish to tell them here that they will be presenting their findings to the class, in Lesson 11.

5. Form teams of students and distribute the materials.

6. Direct each team to evaluate the ideas listed on the board and determine which ideas they think could be tested easily in the classroom. Have them write each testable idea or variable that they could manipulate on a separate card and place them on their Planning Board. (Please see the instructions on pgs. 32 to 33 in the Student Activity Book.) Encourage the teams to plan different experiments so that, together, the whole class can find out more.

7. Now ask students to select one variable to change while keeping the other variables constant. You may need to limit the number of teams that choose each variable or assign teams to test specific variables to ensure that enough materials are available.

8. Now the Student Activity Book instructs students to glue their cards to their team's Planning Board.

9. Next, the students write down the procedure they intend to follow to test the variable with which they will be working. **Activity Sheet 3** will be a helpful guide. It also asks them to list the materials they will need and recommends that they use the "fair test" of magnetic strength from Lesson 4 to find the strength of the electromagnet. You also may want to consider letting students measure electromagnetic strength in a different way, of their own. Keep in mind that this will prevent comparison of the results of different teams.

10. Discuss with the class the need to decide on specific values for the variables that will be held constant, so that the results from different teams can be compared. For example, the teams that are not testing the effect of changing the number of wraps of wire will need to keep the number of wraps at a constant value. Twenty wraps will work. So will 14, 21, and 27, or the number of wraps that can be made with the length of wire provided. But one class value should be chosen so that comparisons can be made.

    Note: To conserve batteries, ask all the teams that are not testing the effect of using different numbers of batteries to use only one battery.

11. Ask the students to draw a picture of the Planning Board in their notebooks. Then have each team prepare their Planning Board sheet for display in the classroom.

12. Ask the members of each team to discuss their plan among themselves and decide how each member will participate in all of the tasks that the team needs to perform.

LESSON 9

Final Activities

1. Ask each student to complete their own **Activity Sheet 3** to reflect their team's plan. You may want to assign this task as homework.
2. Ask the students to write about the procedure they intend to follow and what they think the outcome of their experiment will be. You also may want to ask them to write about what they think the outcome of the other teams' experiments will be.

Extensions

Ask each team to prepare a larger version of their Planning Board for display. Have them include an explanation of the various parts of the Planning Board and a labeled diagram of the experiment.

Evaluation

The Planning Boards are a natural way to evaluate whether the teams understand what they are doing. It is more difficult to assess what the individual students understand, but **Activity Sheet 3** will give some indication.

Reading Selection

Who Is This Story About?

This is a true story about a real person who lived in the 1800s. I am not going to tell you whether this mystery person was a woman or a man, or what this person looked like. I am going to tell you that this person built an electromagnet that could hold 2,000 pounds—1 ton! For now, we will call this person "JH." As I tell you more, think about how you could find out who "JH" is.

JH was born to a family that did not have much money. They lived in Albany, New York. While still young, JH read a book about science that changed JH's life. JH decided to learn as much as possible about science. In the 1800s, people had to pay to go to school, so JH worked as a tutor to pay for JH's education.

JH was a curious person, often taking time to think about and experiment on interesting questions such as: "Why does snow melt faster on a walkway than on grass?" "Why does light change color when it passes though syrup?" and "Does hot water evaporate faster than cold water?"

This person grew up to become a teacher, an important scientist, and then the first secretary of the Smithsonian Institution in Washington, D.C. JH experimented with electricity and magnetism, built some of the very first motors and electric generators, made careful observations and measurements of sunspots, and helped to build the first telegraph.

When JH was experimenting with electricity, there were no easy ways to measure "how much" electricity was flowing through a circuit. So JH and other scientists used their own bodies to measure how much electricity was present. A shock that could be felt all the way up the arm to the shoulder was considered much stronger than a shock that could be felt to the wrist only. We now know that this is not a very safe way to experiment with electricity!

JH never rode in fancy cars, never appeared in television commercials, or sang at rock concerts; all these things were invented after JH died. But this person became famous without doing these things. JH became famous simply by working hard and by trying out ideas until JH discovered something that worked.

Who is this story about?

How can you find out?

LESSON 9

Outlining the Team's Experiment   Activity Sheet 3

NAME: _____

DATE: _____

1. The question (hypothesis) we will try to answer is: _____
   _____
   _____

2. The one variable we will test is: _____

3. In order to make our experiment a fair test, we will keep all of these variables constant (unchanged):

   1) _____
   2) _____
   3) _____
   4) _____
   5) _____

4. Will we need special materials or equipment? _____ If so, we will need:
   _____
   _____

5. What we will measure: _____
   _____
   _____

6. What we will count: _____
   _____
   _____

7. What we will observe: _____
   _____
   _____

8. What we think will happen to our hypothesis: _____
   _____
   _____

LESSON 9

Planning Board   **Activity Sheet 4**

Name: _____

Date: _____

Our hypothesis:

The variable we will test:

The variables that must not change:

How we will test:

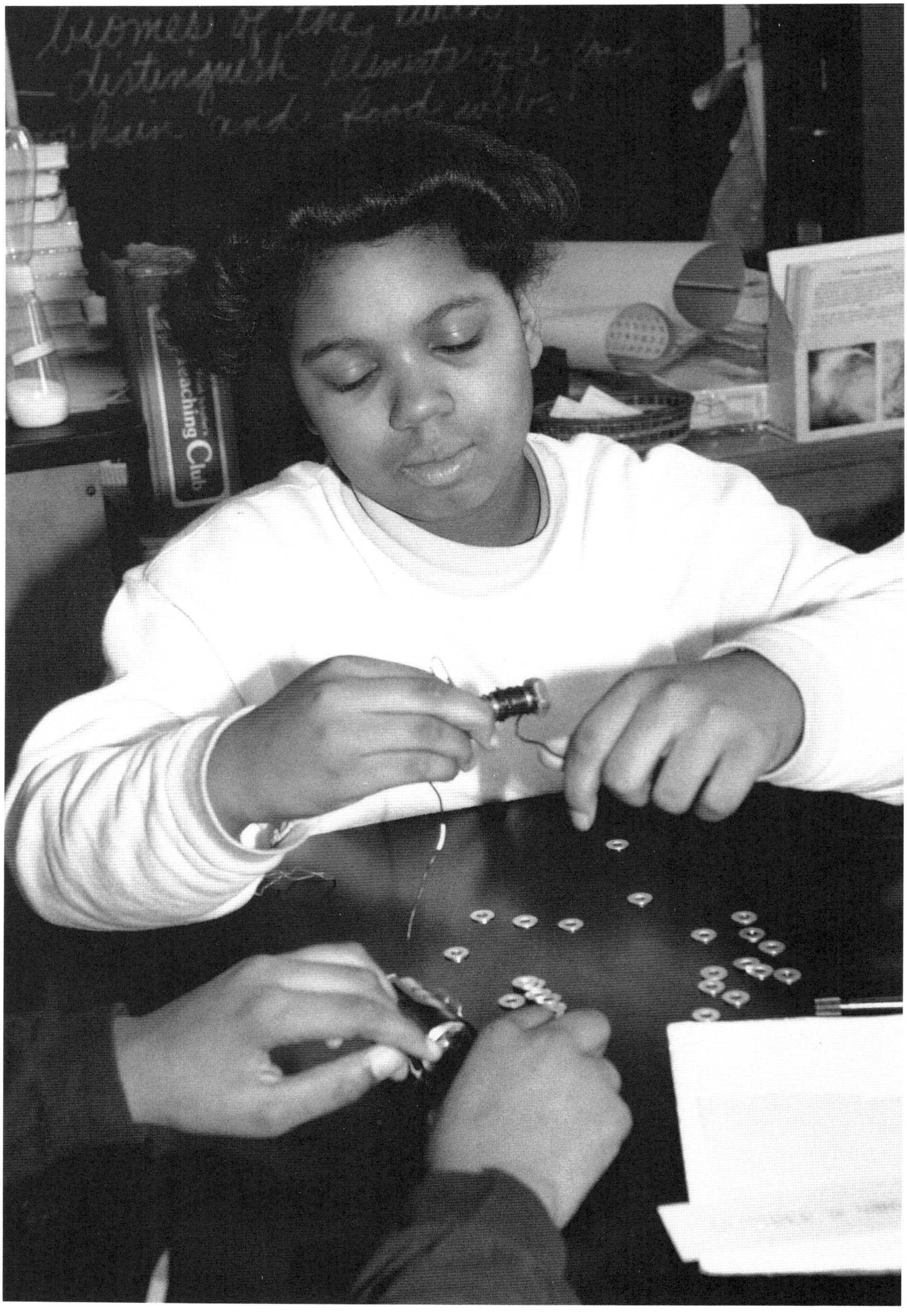

LESSON 10

Testing an Electromagnet

Overview
In this lesson, the teams assemble and conduct the experiment that they designed in Lesson 9.

Objectives
- Students learn how to carry out the experiment that they planned in Lesson 9.
- Students learn as part of a team.

Background
Depending on the ideas that your students had in Lesson 9, you probably will have teams conducting anywhere from two to five different experiments during this lesson. All of the experiments will produce results. The results may not be what you or the students expect them to be, but that is what science is all about.

The point of doing an experiment is to find out what happens and to try and learn from it. Students should be encouraged to experiment by changing only the variable they have chosen and to pay close attention to what happens. Much can be learned from all kinds of results.

Note: Please remind students again **not** to leave the switches on for too long or the wires will get warm and the batteries will weaken.

Materials
The materials for this lesson will vary for each team, depending on the experiment that they are conducting. The "Experimental Materials" list below is meant to cover most of these materials, and corresponds to the categories listed in the **Background** section of Lesson 9. All teams will use some of the same materials, however—see the "For every four students" **Materials** list below.

For each student
1 student notebook

For every four students
1 switch
1 set of 50 washers, USS standard No. 10

LESSON 10

    1 jumbo paper clip
    1 battery and battery holder
    1 piece of #22 coated hook-up wire, 20 cm (8 inches) long
    1 piece of #22 coated hook-up wire, 80 cm (32 inches) long
    1 steel bolt, 8 cm x 6 mm (3" x ¼")
    1 set of experimental materials:
        2 sets of 3 pieces of #22 coated hook-up wire 40 cm, 60 cm, and 100 cm (16 inches, 24 inches, and 40 inches) long
        2 sets of 3 steel bolts, 8 cm (3 inches) long with diameters of 4 mm, 8 mm, and 12 mm (⅛ inch, ⅜ inch, ½ inch)
        2 sets of 3 steel bolts, 6 mm (¼ inch) diameter with lengths of 3.5 cm, 10 cm, and 13 cm (1¼ inches, 4 inches, 5 inches)
        2 sets of 3 different core materials (1 pencil, 1 brass bolt, and 1 aluminum nail)
        2 sets of 3 batteries and battery holders and 2 pieces of #22 coated hook-up wire, 20 cm (8 inches) long

Note: Students may need all or only some of these experimental materials, depending on the experiments they have planned.

Preparation

1. Check the Planning Boards from Lesson 9 to be sure that you have the materials each team will need.
2. Prepare the materials for distribution.
3. If you have not conducted this lesson before, you may want to pick one or two experiments and try them out yourself.

Procedure

1. Begin by reminding the students of the question they are investigating: How can you make an electromagnet weaker or stronger? Discuss with them their various plans for experimenting with an electromagnet. Ask them what they think the various experiments might show. The Student Activity Book also asks these questions.
2. Review with the students the tasks that they will share as they work together in their teams: hooking up the circuit, recording the data, wrapping the electromagnet, and measuring the strength of the electromagnet.
3. Distribute the materials to the teams.
4. While the students conduct their experiments, assist as needed but try not to overcoach. Some teams will need reassurance that there is no "right" answer. The point is that students simply observe what happens and record their findings.

    Encourage the students to discuss their discoveries with other members of their team. If appropriate and time allows, they may want to perform the experiment again.

Final Activities   When the students finish, ask them to write in their notebooks what they have done. The Student Activity Book suggests that they pretend they are explaining their experiment to a relative or an absent classmate. This will give the students a chance to think again about what they have found out.

Extensions   If the students are interested in experimenting further, encourage them to try using additional materials, a different variable, or a different way to measure magnetic strength. For example, they could try using more turns of wire, batteries of different sizes, or different kinds of wire.

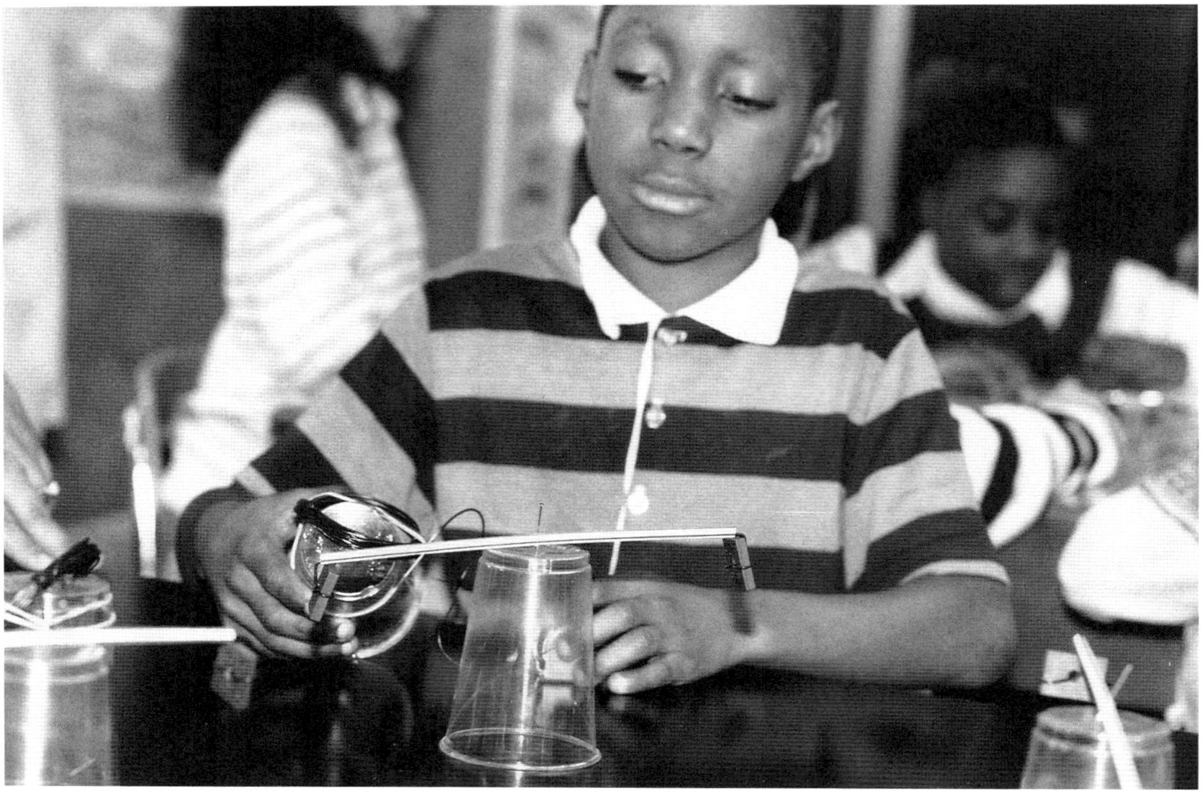

LESSON 11

Showing Others What You Have Learned

Overview

In this lesson, students review the experiment they conducted in Lesson 10. They graph their results and report them to the class. The class compares each team's findings. Students attempt to make connections between what they found out and other students' findings.

Objectives

- Students graph the data they collected in Lesson 10.
- Students discuss their experimental results.
- Students learn how their experiment relates to the work of others.

Background

This lesson gives students the opportunity to communicate what they have learned. It is the occasion for presenting their findings graphically, applying mathematical skills, and making either written or oral presentations of the results of their group's experiment, using language-arts skills. Once each group has presented its results to the class, the students then have time to think about the results and to try to synthesize them into a more coherent view of electromagnetism.

There are many possible ways for students to present their results. Step 2 of the **Procedure** section shows a sample graph. You could show students how to make this kind of graph or challenge them to find other techniques for presenting information.

There are various ways in which students can present the results of their experiments; the choice is yours. They might write up a group report and contribute a copy of it to a bulletin board on which all the reports are displayed. In addition, you might ask each group to make an oral report to the class. As an extra dimension to a presentation, you might suggest that the students pretend they are presenting a paper at a scientific conference. You might challenge them to use visual aids and to adopt the language and style that they imagine scientists use at such conferences. The audience could be challenged to behave appropriately and to try to ask the presenters probing questions.

In whatever way you choose to have the students present their results, the products of their presentations should be kept in a portfolio prepared for each

LESSON 11

student. A portfolio of work will be a useful way to assess the students' progress at the end of the unit.

Materials

For each student
- 1 student notebook
- 1 copy of **Activity Sheet 5** (unlabeled graph paper)

For every four students
- 1 blank grid on a plastic transparency sheet
- 1 transparency marker pen

For the class
- 1 overhead projector and projection screen

Preparation

1. Make one copy of **Activity Sheet 5** (pg. 76) for each student.
2. As you look over the data that the students collected in Lesson 10, you may want to try to identify difficulties that certain teams might have in graphing their data.

Figure 11-1

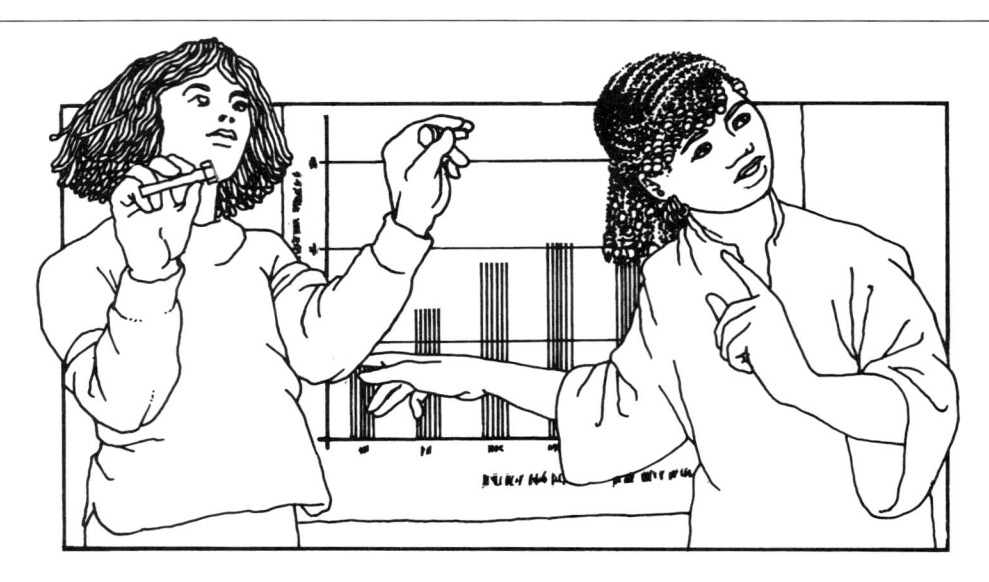

Procedure

1. Begin by explaining to the students that they are going to tell the rest of the class what they learned from their experiment. Describe the purpose of a scientific conference and invite each team to present what they found.

2. Demonstrate the construction of a sample graph on the overhead projector (see the sample data table and graph in Figures 11-2 and 11-3). You also may want to demonstrate the experiment that produced the sample data. Wrap a wire around a permanent magnet and explain to the students that the experiment is intended to see if a different number of turns of wire has an effect on the strength of a permanent magnet. Ask the students to guide you as you go through the steps of graph construction. Why is it important to put a title on it? Why date it? Why label the axes? Why is the scale important?

72 / Showing Others What You Have Learned

LESSON 11

Figure 11-2

Sample data table

The Effect of Wrapping Wire Around a Permanent Magnet on the Magnet's Strength	
Number of Wraps of Wire	Number of Washers Held
0	11
5	11
10	12
15	10
20	11

Figure 11-3

Sample graph

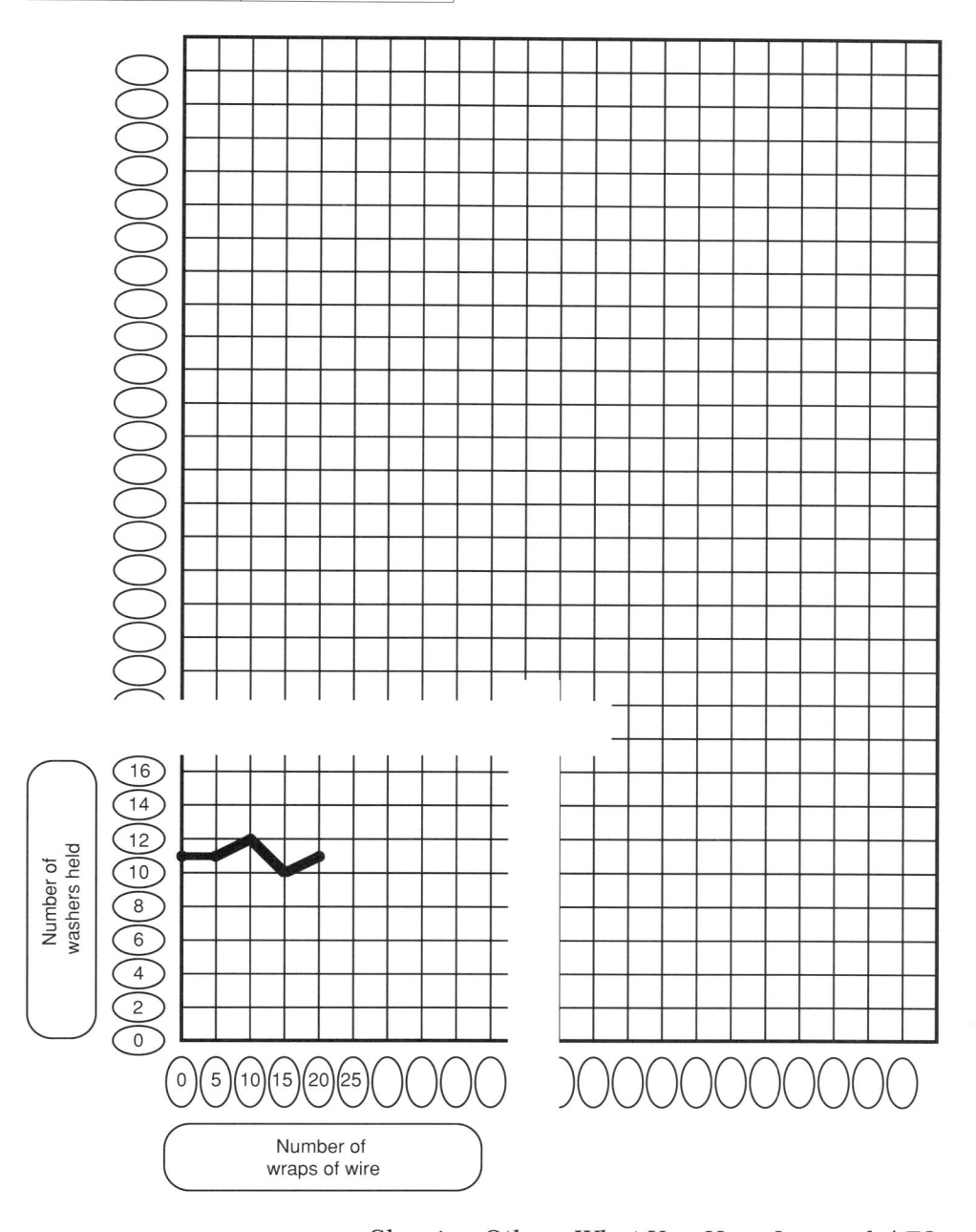

Showing Others What You Have Learned / 73

LESSON 11

3. The data that you are graphing will demonstrate that the number of times a wire is wrapped around a **permanent magnet** does not change its magnetic strength. Ask the students to interpret the graph that you draw and to discuss what they are thinking.

   Some students may have difficulty reading the graph. Help them to see that this is not a "failed experiment," but, rather, a scientifically valuable result. Emphasize the fact that this experiment (with the graphed data as evidence) has determined something important about electromagnetism.

4. Students may be confused by the fact that the number of washers held by the magnet varied slightly during the experiment. This is a good opportunity to talk about the reasons for repeating experiments. Ask them if they can think of reasons why the magnet with 15 wraps of wire held only 10 washers (someone sneezed, one of the washers was bigger than the others, the magnet was bumped). Explain that in all experiments there are some things that it is difficult to control completely.

5. Before beginning the "conference," you may want to discuss the format for presentations with the class. You could (as the Student Activity Book does) ask each team to prepare one graph on the transparency and to be ready to discuss what they thought would happen (hypothesis) and what they learned (findings). You also may want to expand the time allotted for this task to allow for more elaborate presentations. (There are a number of language skills involved in the preparation of presentations. Perhaps this could be combined with a language-arts lesson.)

   Emphasize once again that it is not important that the hypothesis match the findings.

6. Distribute the materials and have the students graph their data from Lesson 10 on their team's transparency. Have each student graph the data on his or her own sheet of graph paper and place the graph in their student notebook.

7. Ask them to write a sentence or two in their notebooks about what the graph shows.

   Remind students that their notebook is a record of what they have discovered.

8. Have each group present its graph to the class. Lead the discussion so that other teams interpret the graphed data. Ask one of the presenting students what his or her team's hypothesis was and then ask another student (on a different team) what he or she would conclude from the graph being presented. You can involve a number of students each time by asking for second and third opinions.

Final Activities

After all of the groups have presented their findings, help the class to summarize the various reports. Ask them how they would use the information gathered to build a super-strong magnet. Encourage the students to discuss their ideas in small groups first, so that everyone has the opportunity to be heard. If you wish, each group could then summarize its discussion for the whole class.

LESSON 11

Extensions Ask students to write a story and draw a picture describing how they would build an electromagnet strong enough to lift a refrigerator.

Evaluation
1. The graphs and the written accounts in the students' notebooks provide a record of their work and are a measure of their level of understanding of the skills and knowledge acquired during Lessons 9, 10, and 11. The graphs and written reports should be retained for the students' portfolios; they are a record of progress and accomplishment.

2. If the students present oral reports, you might invite other classes or parents to attend the presentation. You also might ask students to plan for a later presentation if you choose to do a final presentation of results after the class has completed the unit.

3. Students' presentations offer an excellent opportunity to coach students on presentation style as well as to give them experience speaking in front of a group. They also provide opportunities for the class to listen carefully and to develop sharp but courteous questioning techniques.

4. Encourage the students to write about their graph and to explain what they learned from it. Urge them to explain the experiment they performed and to describe what they learned from it. You may want to suggest that they make some copies of the unlabeled graph paper (**Activity Sheet 5**) from their Student Activity Book in case they conduct similar experiments on their own later and want to graph their results.

LESSON 11

Activity Sheet 5

NAME: _____

DATE: _____

| LESSON 12 | **Making a Motor** |

Overview In this lesson, students use an electromagnetic coil and a switch to investigate ways to make the compass they built in Lesson 5 spin like a motor. They review the concept of magnetic attraction and repulsion and examine once again the usefulness of on/off switching. Both of these ideas are needed to understand the electric motors that students will experiment with in Lessons 13 and 14.

Objectives
- Students learn how to use electromagnetism to make a compass rotate.
- Students begin to understand how a motor functions.
- Students express their ideas about how motors work.

Background The electric motor is used in a great number of modern electric appliances. Motors use electromagnetism to convert electricity into mechanical work. The work that motors perform is often work that people used to perform.

The idea behind an electric motor is straightforward. The basic ingredients are electricity, magnetism, and a way to turn the electricity on and off at the right time. By turning on and off the electricity flowing to an electromagnet at the right moment, the electromagnet can be made to attract (when the switch is on and the electricity is flowing), then coast (when the switch is off). However, many commercial motors are designed to switch the direction of the electric current, making the electromagnet alternatively attract, then repel. This is more efficient and enables the motor to run more smoothly. If you or one of your students were to take apart a motor from an electric appliance, you probably would not find any permanent magnets. This is because the motors in most appliances use electromagnets (coils) instead of permanent magnets. These motors use electrical current from a wall outlet to operate, rather than current from a battery. A number of ingenious ways have been devised to switch the electromagnets in motors on and off automatically, but the purpose of the motor is unchanged: to use electricity and magnetism to turn the shaft so that the motor can be used to do work.

The motor in this lesson does not have an automatic switch—the switch is operated at the proper moment by the student. It is constructed from an electromagnetic coil and the straw compass the students constructed in

LESSON 12

Lesson 5. This motor could be used to do a small amount of work, but its real purpose is to demonstrate how a motor works.

Materials

For each student
- 1 student notebook

For every two students
- 1 straw compass (built in Lesson 5)
- 1 coil of #22 coated hook-up wire, 1.5 m (5 feet) long
- 1 plastic cup and lid
- 1 piece of #22 coated hook-up wire, 20 cm (8 inches) long
- 1 battery and battery holder
- 1 switch
- 2 rubber bands, No. 16

Preparation

1. Prepare the materials for distribution. You may want to have some students help cut and coil the 1.5-meter wires around their hands, then strip the insulation off the ends. The wire stripping procedure described in **Appendix E** may again be used.

2. If you have not given this lesson before, you may want to set up a motor for yourself and try out the activities ahead of time.

Procedure

1. Begin by asking the students to think about the ways they made a compass needle move in Lesson 8. Ask, "How do you think you can make the straw compass spin by attraction and repulsion?" As the students discuss their ideas, make a list of their questions and thoughts.

2. Ask students to think about what a spinning compass could be used for. List their questions and ideas. This list may lead students to start thinking of the spinning compass as a motor. Let them make the connection on their own, if possible. But, if you need to help the students make this big step, you might note something like, "The compass can spin like a motor."

3. Distribute the materials and remind students that they will be working with a partner and will need to share materials.

4. Now ask students to turn to pg. 43 in their Student Activity Books (pg. 80 in the Teacher's Guide) and follow the instructions for making a motor with a coil and a compass. In this activity, the students will be deciding how to make the compass spin and how to control the spin. Encourage them to discuss these points with their partner. The activity is summarized below.

   - Students review the original function of a magnetic compass.
   - Students find which end of the compass is attracted by the coil and which end is repelled.
   - Students investigate ways to make the compass (motor) spin by turning the switch on and off at the appropriate time.
   - Students try to find a way to control the direction of the spin by using only the switch.

5. After the students have completed their activities, ask them to dismantle their compasses and return their materials to the storage area.

Final Activities

1. Ask the students to write in their notebooks about what they did to control the spin of the motor. Encourage them to write as if they were explaining how to do this to someone who has not yet tried it.

2. Challenge the students to try to think of a way to turn the switch on and off automatically so that the "motor" will run by itself. This is a difficult problem, but some students may come up with designs that hang wires from the straw or that have the compass bump a switch.

Extensions

1. Invite individual or groups of students to attempt to find ways to increase the speed of rotation of the "motor." Suggestions you may want to offer are:

   - What happens if you change the length of the straw?
   - What happens if you use two batteries?
   - What happens if you use two (or more) coils?
   - What happens if you vary the size of the coil?

2. Invite the students to attempt to build automatic switching devices for the compass on the basis of their design ideas from **Final Activities**, Step 2. Ask them to plan carefully what they will do, then let them try it. Some or all of the designs may not work, but the process is valuable for the students.

LESSON 12

Student Instructions: Can You Make a Motor?

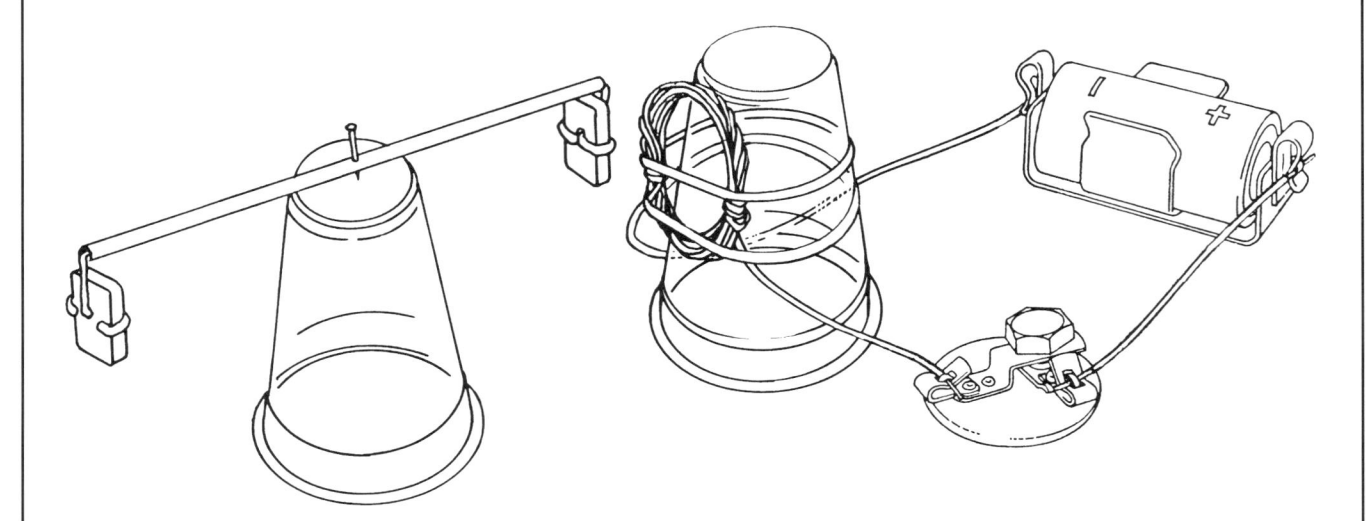

1. Let the compass come to rest so that it is facing north-south. What do you think causes all of the compasses in the room to point in the same direction?

    Using the piece of wire that is 1.5 m long, make a coil as you did in Lesson 8. Attach it to the side of the inverted plastic cup with the two rubber bands. Connect the coil to the battery and the switch. With the 20-cm wire, connect the switch and the battery. Position the coil two finger-widths away from the north pole of the magnet, as shown in the picture above.

2. What effect does the coil have on the magnet when the switch is **off**?

3. Turn the switch on so that electricity will run through the coil. Is the north magnet **attracted** or **repelled** by the coil?

    Which pole of the electromagnet do you think is closer to the compass? What makes you think that? You may want to look back in your notebook at what you learned in Lesson 8.

4. Now, position the south pole magnet next to the coil. Can you predict what will happen when you turn the switch on? Try it!

80 / Making a Motor

5. **Challenge:** Can you make the motor work?

   Try to make the straw compass (motor) spin continuously in one direction just by turning the switch on or off at the right time.

   Remember which end is being attracted to the coil and which end is being repelled. You may want to mark the ends to help you keep track.

   Take turns with your partner. One person can start the motor spinning in one direction, and then the other person can try to make it stop and spin in the **other direction**.

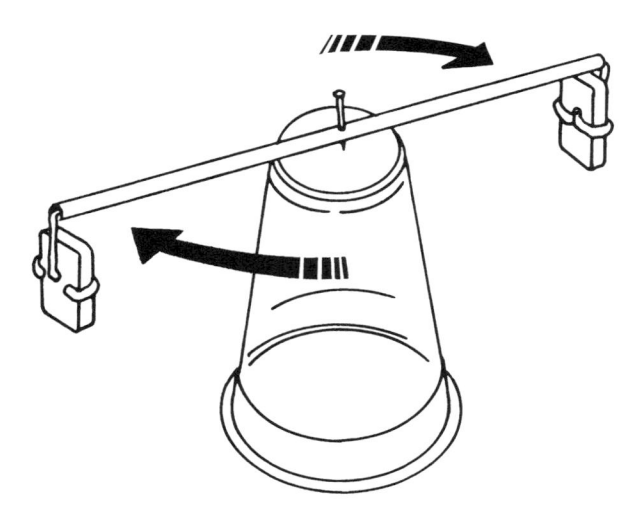

Making a Motor / 81

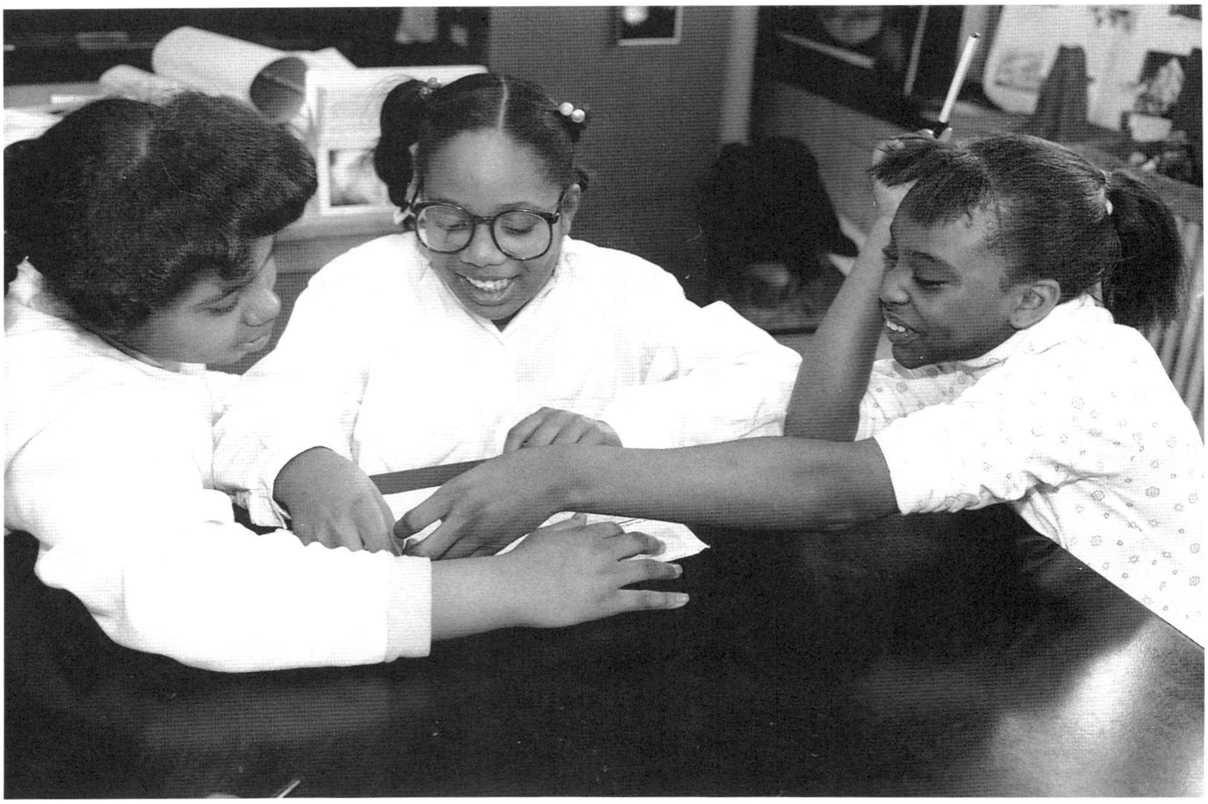

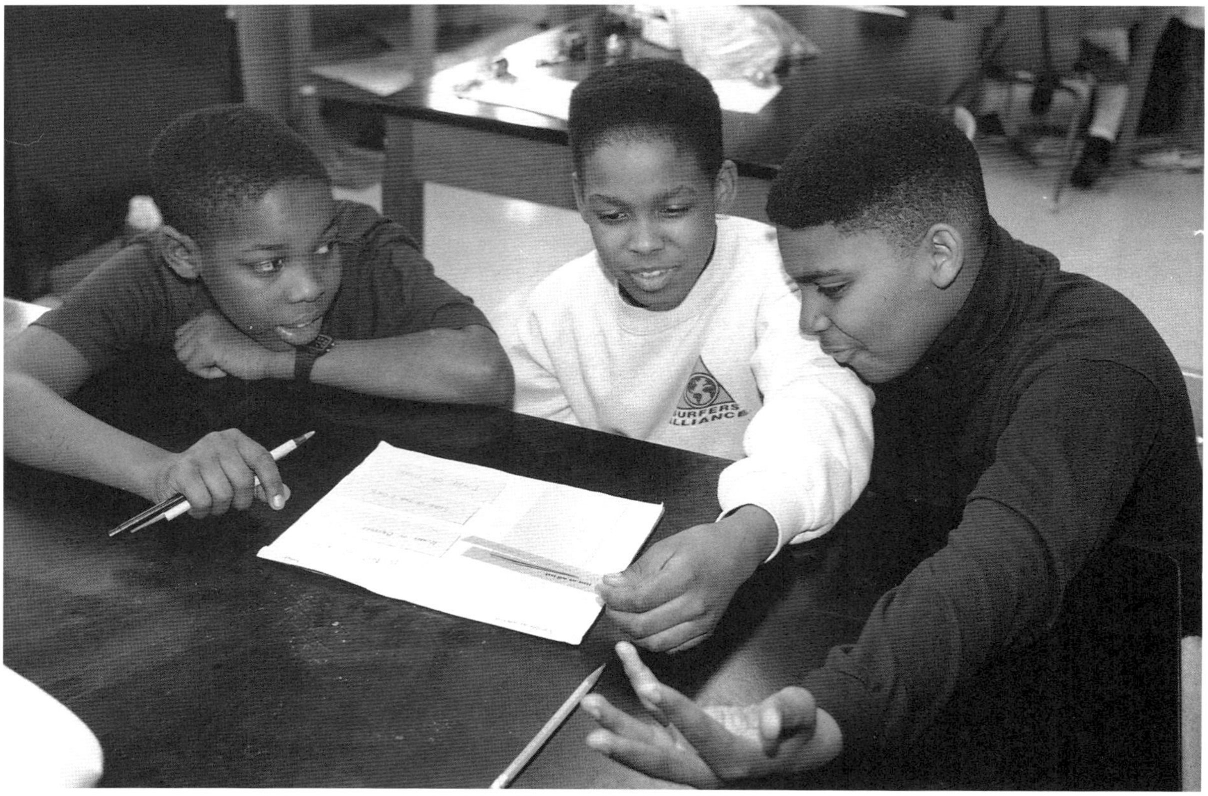

LESSON 13	**Building a Spinning Coil Motor**

Overview	In this lesson, students build a simple, working motor out of inexpensive materials. They learn to troubleshoot and to gain confidence in their manipulative skills. This lesson provides a conceptual transition from the spinning compass motor of Lesson 13 to the more sophisticated motor investigated in Lessons 14 and 15.
Objectives	■ Students build a working electric motor and investigate further how a motor works. ■ Students express their ideas about motors and their uses.
Background	The first working motors were built in the 1830s and performed only limited amounts of work. They were fascinating devices that incorporated automatic switching so that the motor would run by itself as long as electricity was supplied to the electromagnet. Today's motors still operate on the same principles. In Step 2 of the **Final Activities** in Lesson 12, students were challenged to think about ways to switch on and off automatically the electricity flowing through the coil. In this lesson, the motor they build uses automatic switching, but how it works will not be obvious. What happens is that the bouncing of the coil, as it spins, interrupts the flow of electricity and provides an automatic on/off switch that causes the coil to keep spinning. Do not expect all of your students to understand this immediately. One purpose of the next several lessons is for students to develop a concept of how a motor works. You will have an opportunity to assess their thinking about this in **Appendix A**, the Post-Unit Assessments.

Materials	*For each student*
	  1 student notebook
	  2 pieces of #20 bare copper wire, each 20 cm (8 inches) long
	  2 rubber bands, No. 16
	  1 plastic cup and lid
	  2 alligator clips
	  1 battery and battery holder

LESSON 13

    1 switch
    3 pieces of #22 coated hook-up wire, each 20 cm (8 inches) long
    1 piece of #28 enameled copper wire, 65 cm (25 inches) long
    1 piece of sandpaper, 5-cm square (2-inches square)
    1 flexible magnet, 25 x 20 x 5 mm (1" x ¾" x ³⁄₁₆"), with a 5-mm (³⁄₁₆-inch) hole in the center

For the class
    6 screwdrivers

Preparation

1. Prepare the materials for distribution. You may want to enlist the aid of students in cutting the sandpaper and wires into the dimensions specified in the materials list. Information on how to make and repair alligator leads is given in **Appendix F**. This activity could be incorporated into the lesson.

2. If you have not conducted this lesson before, you may want to set up a motor for yourself and try out the activities ahead of time.

Procedure

1. Begin by leading a discussion about physical work. Ask the students to tell the class about the hardest physical work they have ever done. How could an electric motor be used to help do some kinds of hard work? What else are motors used for? Make a list of the students' ideas.

2. Next, distribute the materials to the students and ask them to follow the instructions for building a working motor on pgs. 47 through 49 in the Student Activity Book (pgs. 86 through 88 of the Teacher's Guide). Encourage the students to be patient and to keep trying. Encourage students to take some time to figure out how the electricity gets switched on and off automatically. Many students will be challenged by this activity. Resist the temptation to build the motor for them. Offer suggestions, such as "Bend the ends of the wire a little more" or "Bend this part so that it will not bump into the wire."

3. As students work, the Student Activity Book asks them to think about the following questions and to write about them in their notebooks:

   - How is the electricity switched on and off automatically?
   - Where do you think the electricity flows in this circuit? Draw a picture in your notebook to help you explain.
   - Do you think the coil is magnetic? Why do you think that?
   - How can you make the coil spin faster or change direction? Try out your ideas!
   - How does this motor compare to the spinning compass motor? How are they alike? How are they different?

4. Encourage the students to talk with each other about what they are observing. Ask those who are successful quickly to share their ideas with students who request help, but caution them not to build the motor for them.

84 / Building a Spinning Coil Motor

5. Before asking the students to dismantle their motors and return the materials to the storage area, choose one assembled motor to save for use as a demonstration for **Appendix A**.

Final Activities Tell students that in the next lesson they will be taking apart a different kind of motor that someone else put together.

Student Instructions for Building a Motor

Use the directions and the pictures on these two pages to build an electric motor. If the motor does not work, try changing one thing at a time until it operates. That way, you will know what the trouble was. As you learned in Lesson 5, this kind of problem-solving is sometimes called **troubleshooting.**

1. Start by making a loop in the middle of each of the two bare copper wires. Wrap them around a pencil, then slide the loop off the pencil.

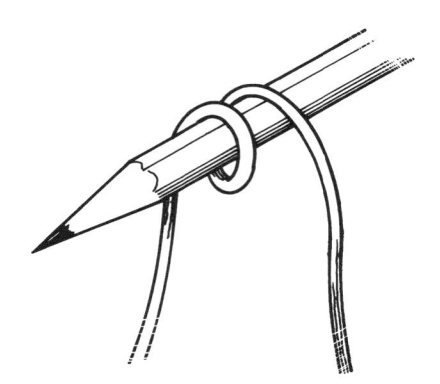

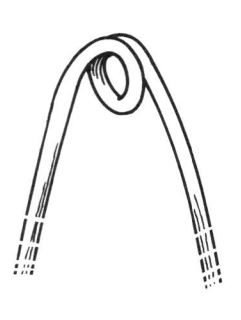

2. Use the two rubber bands to attach the bare copper wires to the plastic cup.

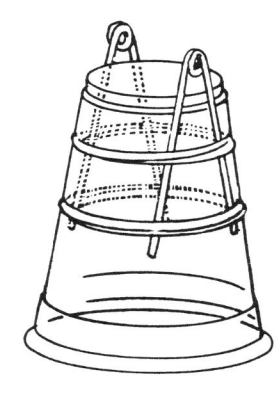

3. Clamp an alligator clip to one end of each of the bare copper wires.

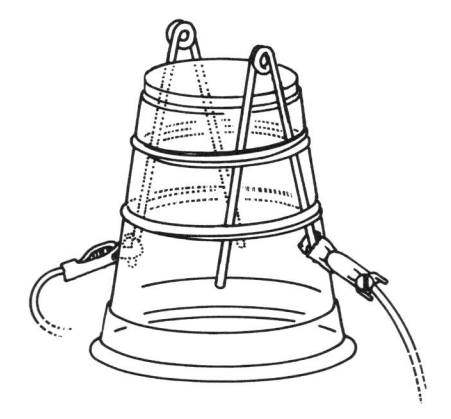

86 / Building a Spinning Coil Motor

4. Hook up the rest of the circuit by connecting the battery and the switch to the alligator clips with the three pieces of #22 wire. **Appendix F** tells you how to make alligator leads.

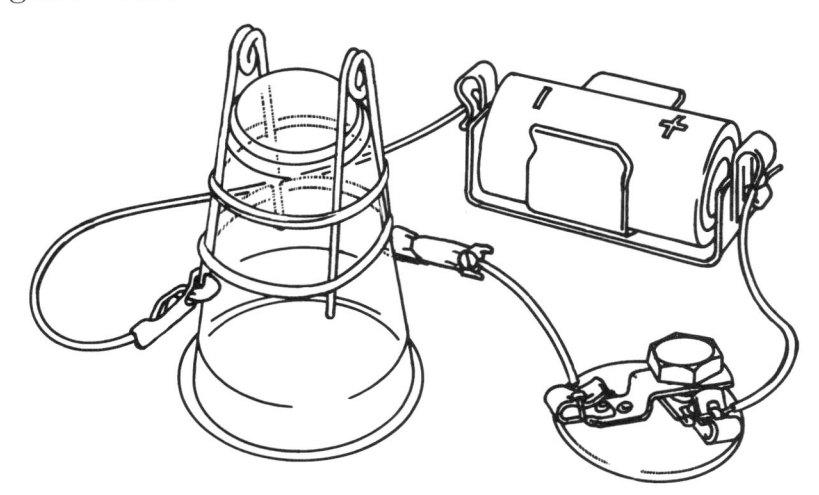

5. Next, begin to make the moving coil from the #28 wire. The wire has a thin coating of insulation on it called enamel, so you will need to sand off the insulation at the ends. This is so that the electricity can flow from the bare copper wire through the thin wire to the other bare copper wire.

   Sand about one finger-length of insulation gently off each end of the wire. You can do this by folding the sandpaper in half and pulling the ends of the wire carefully between the two sides of the sandpaper.

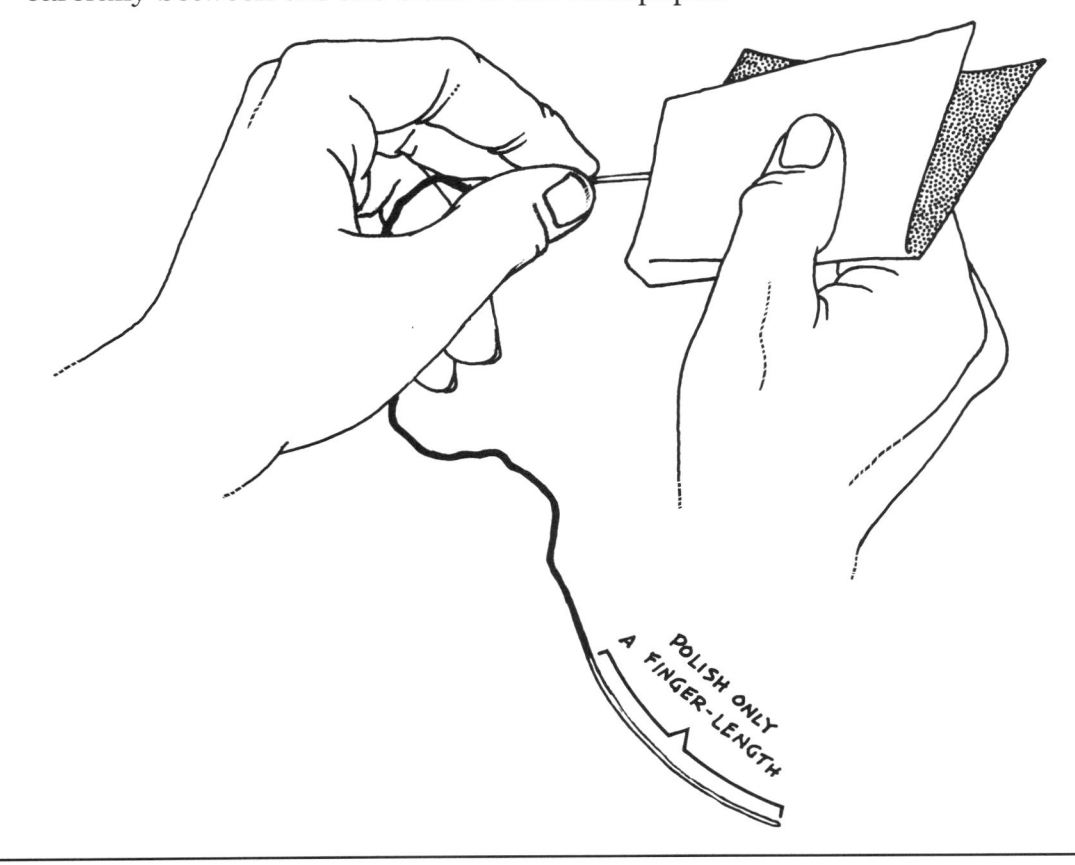

POLISH ONLY A FINGER-LENGTH

Building a Spinning Coil Motor / 87

6. Now, wrap the wire around the battery several times. Leave the bare ends sticking out of the coil. Slip the coiled wire off the battery. Wrap the bare ends three or four times around the coil, to hold it in a circular shape. Then bend the ends of the wire so that they stick straight out on opposite sides of the coil.

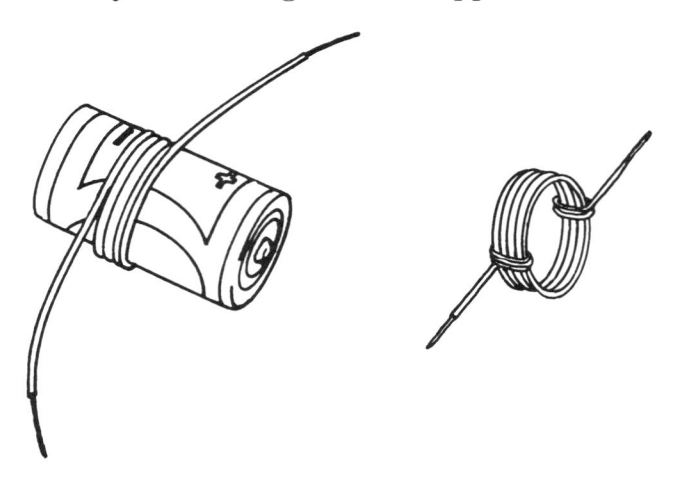

7. Next, place the ends of the coil through the loops in the bare wire on each side of the cup. If necessary, bend the coil and adjust the loops so that the coil can spin freely, without hitting anything.

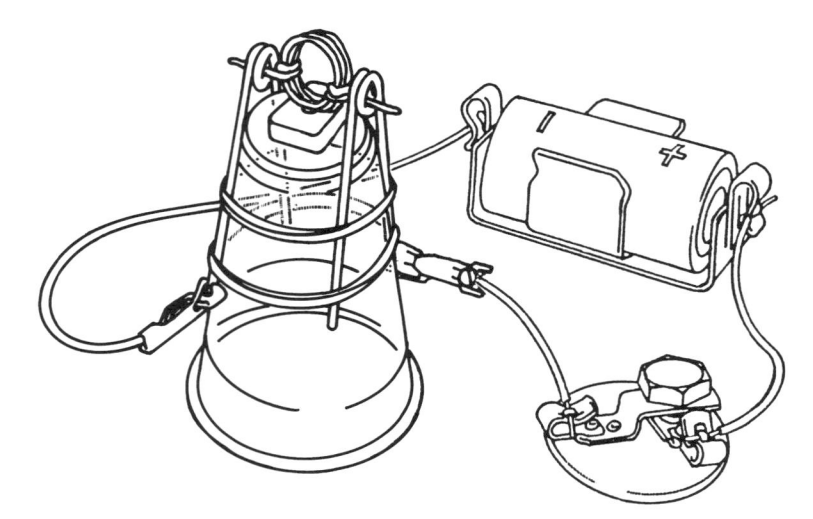

8. Place a magnet on the top of the inverted cup, underneath the coil. Turn the switch on and blow gently on the coil to help get it started.

   If the coil will not spin continuously, try putting the magnet somewhere else, turning it over, or bending a few wires a little.

| LESSON 14 | **What Is Inside an Electric Motor?** |

Overview

In the last two lessons, students built two different kinds of motors; now, they will investigate small, commercially produced motors. Each student makes the motor operate and then takes it apart. This lesson is the link that joins the magnet investigations begun in Lesson 1 with the electromagnet and motor lessons.

Objectives

- Students demonstrate how to operate a small electric motor.
- Students learn what is inside a small electric motor.

Background

People, particularly young people, seem to have an innate curiosity about how things work. From a baby's rattle to a computer, children want to look inside and see what is going on. An electric motor is no exception, but, until the advent of inexpensive miniature motors, it was impractical for large numbers of students to dismantle one. For this reason, few people understand how a motor works or what is inside it.

The motor that students experiment with in this lesson operates on current from a battery. It uses two permanent magnets (attached to the inside of the case) and three electromagnets (wound around the rotor, which is also called the **armature**).

The automatic switching that was discussed in Lessons 12 and 13 is performed by the motor's **brushes** and **commutator**. See Figures 14-1 and 14-2 for two different views of a motor.

As the magnets and electromagnets attract and repel to rotate the armature, the brushes make contact with different segments of the commutator. These segments are connected to the three electromagnets. As the motor rotates, the brushes touching different commutator segments cause the three electromagnets to be turned on and off continuously. Figure 14-3 shows how the motor rotates.

LESSON 14

Figure 14-1

Three-quarter view of a motor

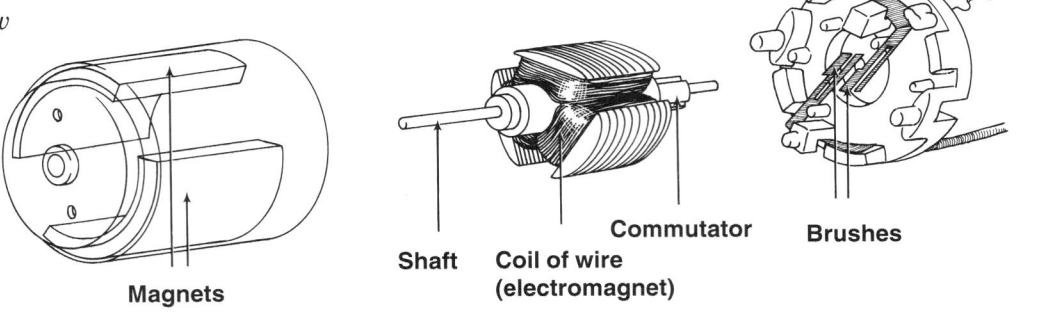

Magnets
Shaft
Coil of wire (electromagnet)
Commutator
Brushes

Figure 14-2

A cross-section view of a motor

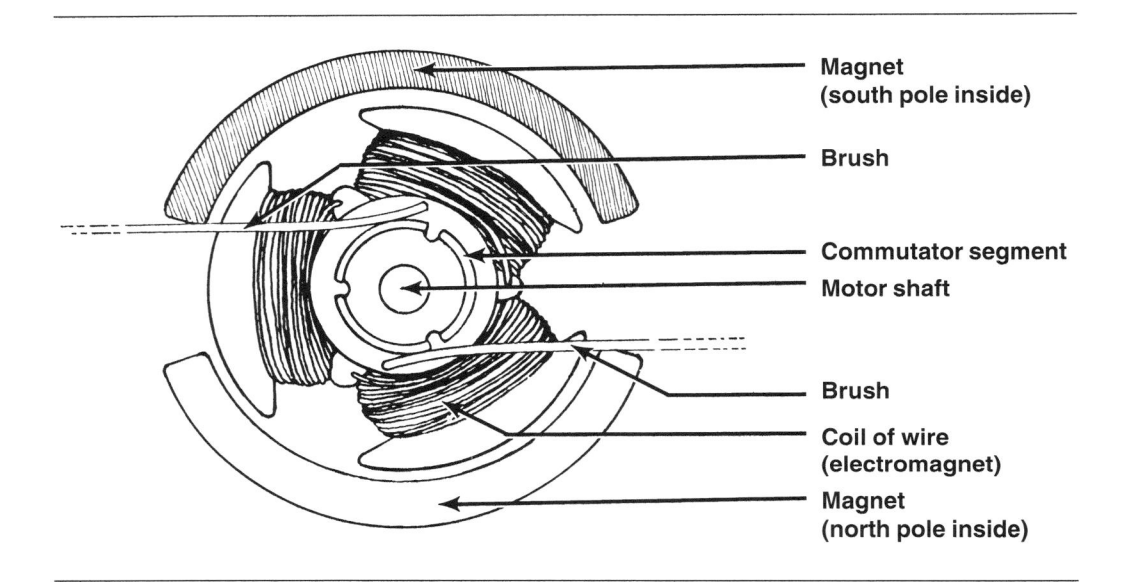

Magnet (south pole inside)
Brush
Commutator segment
Motor shaft
Brush
Coil of wire (electromagnet)
Magnet (north pole inside)

Figure 14-3

How a motor rotates

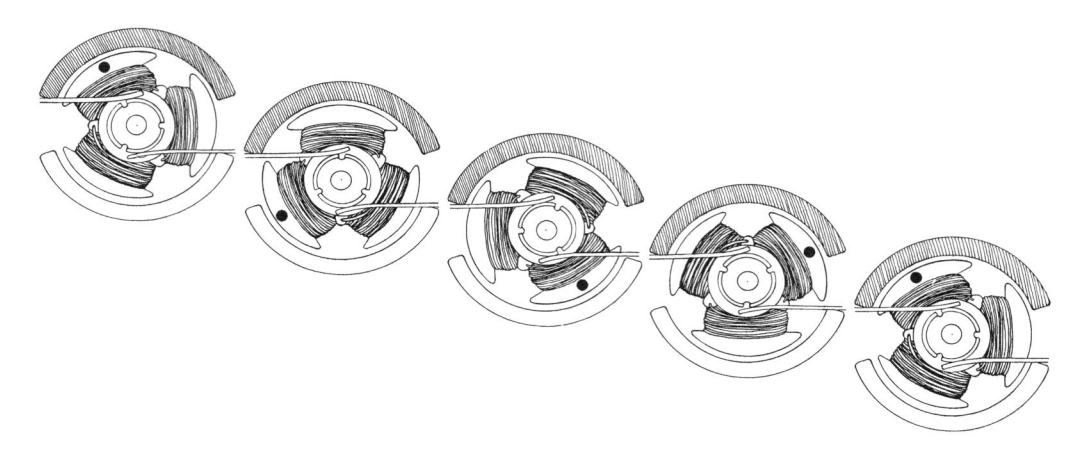

The direction of the electric current through the wire coils also switches each time that the brushes move to a new pair of commutator segments. Changing the direction of the current changes the **polarity,** or "where the poles are." The north and south magnetic poles change places. This allows the

90 / What Is Inside an Electric Motor?

electromagnets and the permanent magnets to attract and repel, depending on where the electromagnet is in its travel around the inside of the motor. Automatic commutator switching is an extremely clever technique that took years to develop. Do not expect the students to grasp all of its intricacies right away.

Materials

For each student
- 1 student notebook
- 1 battery and battery holder
- 1 small electric motor
- 1 steel nail
- 1 plastic drinking straw
- 1 switch
- 1 piece of #22 coated hook-up wire, 20 cm (8 inches) long

For the class
- 1 set of demonstration materials (as in Lesson 2):
   - 1 small, electric motor with wire leads, 1 plastic drinking straw, 1 battery and battery holder
   - OR
   - 1 student- or teacher-supplied motorized toy (with batteries)

Preparation

1. Prepare the materials for distribution. Make sure that the ends of the wire motor leads are stripped and the exposed strands twisted. Student helpers could be recruited for this task.

2. You may wish to dismantle partially the device you will be demonstrating when you begin the lesson.

Procedure

1. Show the class a device that uses a small electric motor. Dismantle the device to the point where the motor is clearly visible, then ask, "What do you think is inside?"

   Allow the students to discuss ways to find out what is inside. Then tell them that they will get a chance to take apart a small motor to find out for themselves.

2. Distribute the materials and ask the students to attach the straw to the shaft of the motor using the directions in the Student Activity Book on pg. 51. They will make a tiny hole on each side of the straw with a nail. Then they will push the motor shaft through the hole. This will allow them to tell which way the motor is turning. Now, ask them to try to use the switch to make the motor start and stop. Challenge them to find a way to make the motor run in both directions (by turning the battery around in its holder or connecting the wires to the opposite ends of the battery holder).

   Encourage the students to talk with one another about what they are doing. Many students will want to show you what they have accomplished.

What Is Inside an Electric Motor? / **91**

LESSON 14

If, after five minutes of experimenting, some students are still struggling with the task, ask one of the students who has had success to tell the class how it was done. Allow time for the students who needed help to try these suggestions.

3. Tell students to try to hook up an electric circuit that includes a switch. Ask them to try to make the motor stop and start by using the switch.

4. Next, tell students to try to figure out a way to make the motor turn in the opposite direction. Encourage them to talk among themselves about how they did it.

5. Next, ask the students to follow the directions for taking the motor apart on pgs. 52, 53, and 54 in the Student Activity Book. These directions lead students to:

   - dismantle the motor by bending back the metal tabs with a nail
   - look closely at the parts of the motor and how they fit together
   - try to determine what the parts do

6. Ask students to look at their motors and at the three-quarters view of a motor (Figure 14-1), which is in the Student Activity Book on pg. 53. Ask them to think about what the different parts do and why. Ask them to write their ideas in their notebooks.

7. Ask the students to leave the motors disassembled. They will put them back together at the end of the next lesson. Have the students return the materials to the storage area.

Final Activities

1. Give the students time to think about how to put the motor back together. Suggest that they write their plans in their notebooks.

2. Give the students an opportunity to discuss the purposes of the various parts of the motor. Tell them the names of the various parts but do not overemphasize the need to learn the "right" names for the parts.

Extensions

1. As a language lesson, discuss how various objects get their names. Have the students research the origin of words. Ask them to invent replacement words for words that are hard to remember. For example, "wind speed indicator" might replace "anemometer." What makes certain words easier to remember than others?

2. In this lesson, the students learned what is inside a motor by taking one apart. Ask the students to think of other things that they could take apart (with the owner's permission) in order to learn how they work. What are some things that cannot be taken apart? What are some things that are still mysteries after you take them apart? Allow the students to discuss their ideas and wonder about things they have neither seen nor done.

Evaluation

These small motors operate very efficiently, but they are not as easy to understand as the motors that the students built themselves in Lessons 12 and 13.

Some students will quickly connect the functions of the magnets and electromagnets in this lesson with their functions in previous lessons. Other students may not recognize that the different types of motors have anything in common. Most students will have some difficulty understanding the role of the brushes and commutator in switching the electricity automatically. Do not be concerned about all of the students making these connections now. Lessons 15 and 16 provide additional occasions for them to develop these ideas.

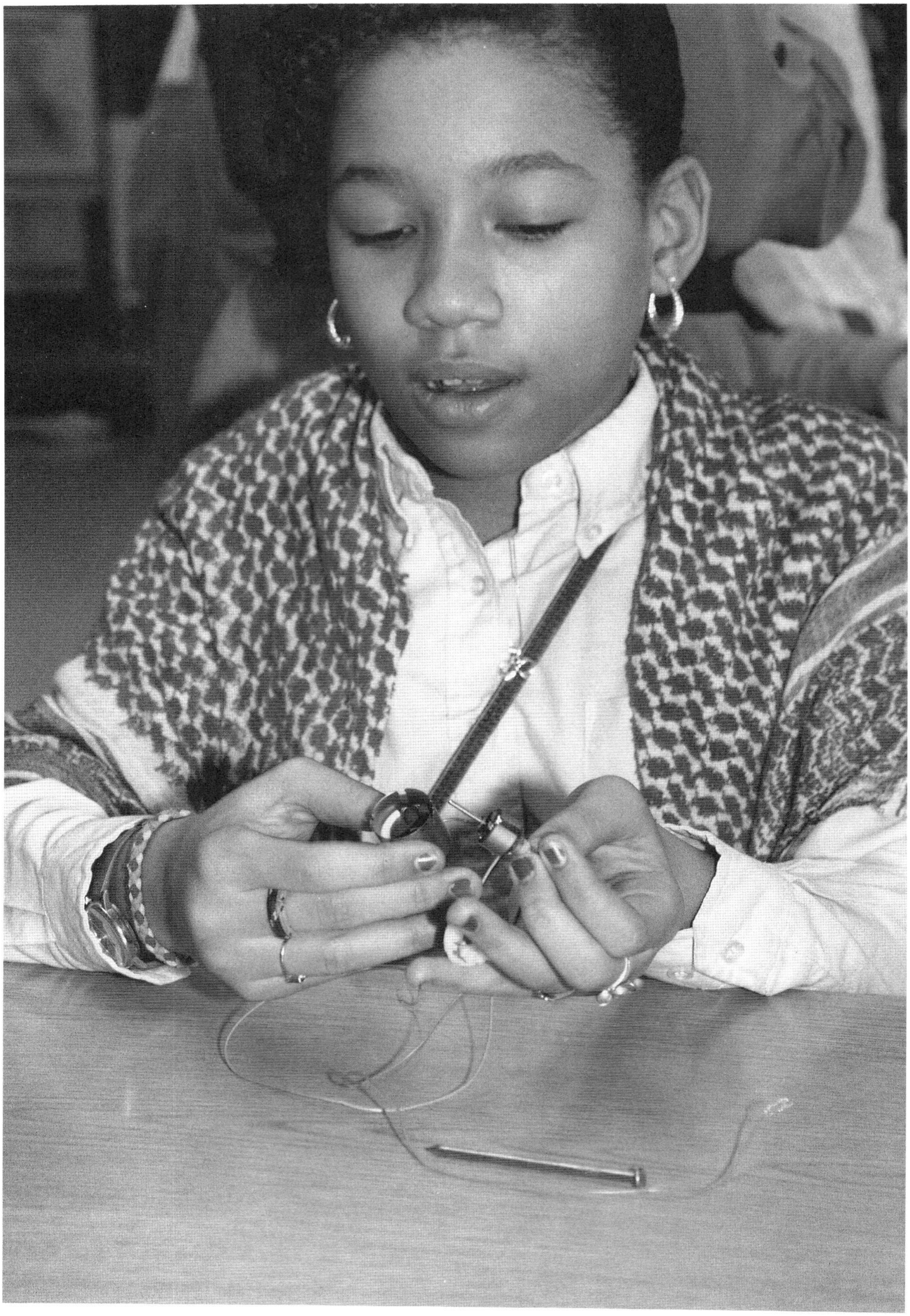

| LESSON 15 | **How Does a Motor Work?** |

Overview

Now that the students have learned what is inside a motor, they use the armature of the motor they disassembled in Lesson 14 to experiment with variables that have an effect on the motor's operation. They use what they have learned to reassemble the motor and test to ensure that it is working properly.

Objectives

- Students explore different ways of changing how a motor functions.
- Students learn how a motor works.
- Students successfully reassemble the motor they took apart in Lesson 14.

Background

The purpose of taking apart the motor in Lesson 14 was to see what is inside it. In this lesson, students discover how all those parts work together.

At this point, students know that a motor needs electricity, magnetism, and a special switch in order to function. Now, they are provided with an opportunity to experiment with these elements.

In this lesson, the brushes of the motor will be two wires that make direct contact with the commutator. The students will use the armature that they removed from the motor. The magnets are the same as those they have been using since Lesson 2. All of these items will be assembled on a "test stand"—where the parts can be seen and manipulated easily. This setup is shown in Figure 15-1.

There are several variables that students can change that have an effect on the operation of the motor:

- the location of the magnets (above, below, or to the side of the armature)
- the proximity of the magnets to the armature
- the poles of the magnets facing the armature (if like poles face the armature on opposite sides, the motor will not turn; also, the direction of spin can be controlled by reversing the poles)
- the number of batteries providing electricity to the electromagnets on the armature

How Does a Motor Work? / **95**

LESSON 15

- the direction in which the electric current flows (this can be changed by flipping the battery over in its holder)
- the presence or absence of electric current (achieved by turning a switch on or off or by disconnecting the circuit).

By exploring ways to control the motor, the students develop for themselves a concept of how a motor works. They then use what they have learned to put the motor back together and to test to see that it is working.

Figure 15-1

The commutator must rest in the fork of the V-shaped wire or brushes

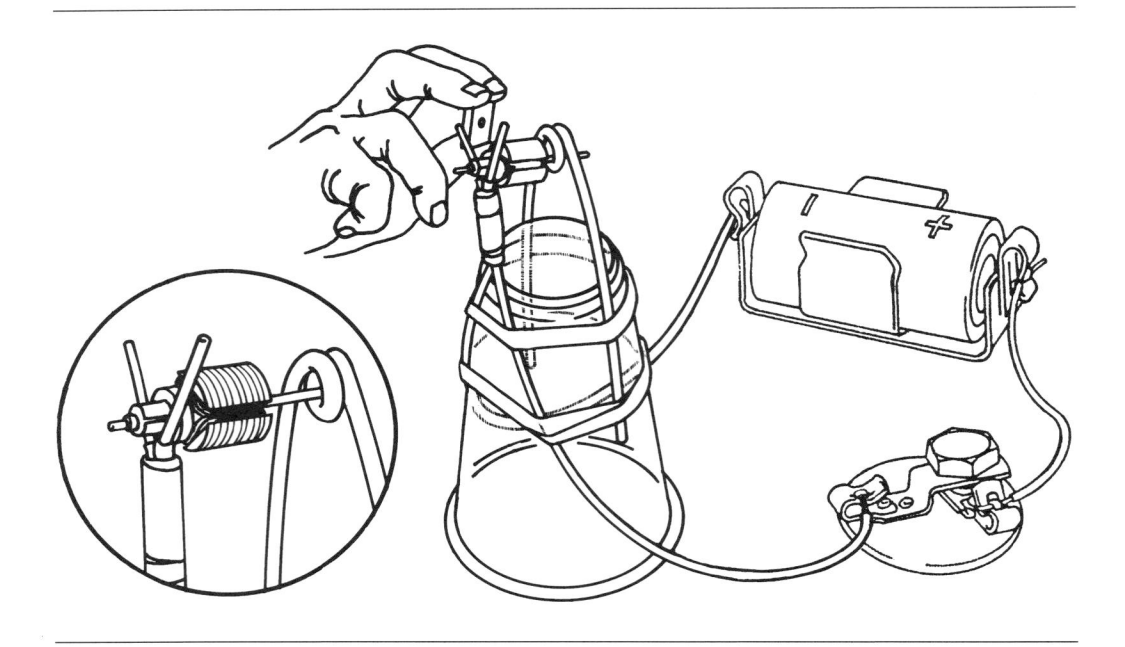

Materials

For each student
- 1 student notebook
- 1 disassembled motor (from Lesson 14)
- 2 rubber bands, No. 16
- 2 flexible magnets, 25 x 20 x 5 mm (1" x ¾" x ³⁄₁₆"), with a 5-mm (³⁄₁₆-inch) hole in the center
- 1 double strand of #20 coated hook-up wire, 30 cm (12 inches) long
- 1 piece of #20 bare copper wire, 20 cm (8 inches) long
- 1 piece of #22 coated hook-up wire, 20 cm (8 inches) long
- 1 battery and battery holder
- 1 plastic drinking straw
- 1 switch
- 1 nail, common 12D

For the class
- 1 roll of PVC electrical tape

Preparation

1. Prepare the materials for distribution.

    Note: If this is the first time that the materials have been used, you may have to construct double strands of hook-up wire from the 2 pieces you have for every student. Students could help with this task. Strip both ends of two pieces of #20 wire (each 30 cm, 12 inches, long) and fasten the wires together with electrical tape, as shown below. The tape should hold the wires together where the insulation ends and the bare wire begins.

Figure 15-2

Making double-strand hook-up wire

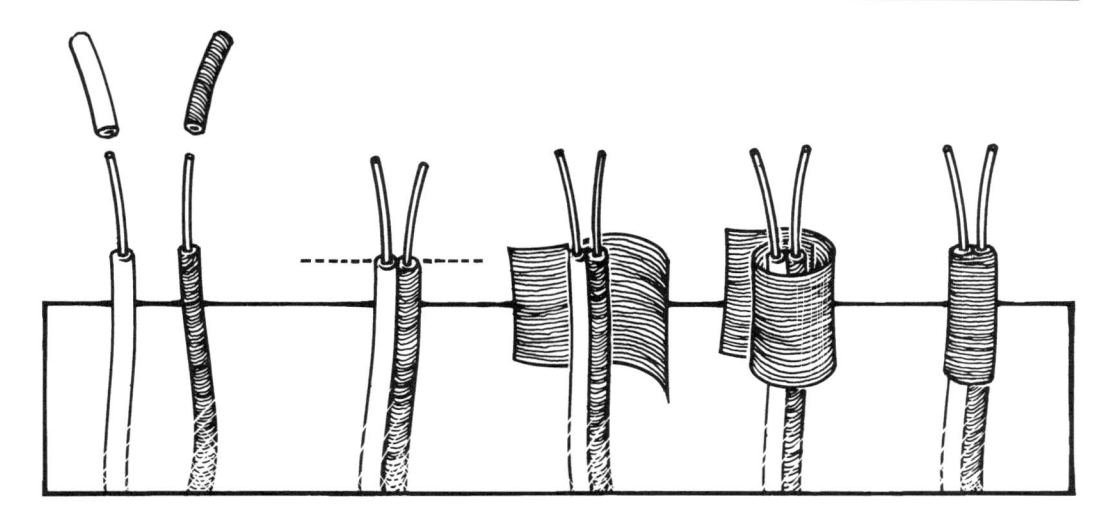

Procedure

1. Ask the students to remind you of the basic components needed to make a motor run. Briefly review with them what the different parts of the motor do by asking them to discuss what they remember from previous lessons.

2. Explain to the students that they are going to try to make the armature spin outside the motor case—where they can see it work. Tell them that, next, they will be experimenting with ways to make the armature turn in different directions and at different speeds. They will be trying to fathom how a motor works. Finally, they will see if they can put the motor back together and make it run again.

3. Distribute the materials to each student and ask them to follow the directions on pgs. 55 through 58 in their Student Activity Book to help them experiment with the parts of the motor.

   Students will be asked to do the following:

   ■ Set up their test stand, then experiment with the magnets to find the location that makes the motor spin, then the location that makes the motor spin the fastest.

   ■ Experiment to find as many ways as possible to make the motor turn slowly.

   ■ Experiment to find as many ways as possible to make the motor change direction.

   ■ Think about how the parts fit together and follow suggestions for reassembling the motor and testing it.

   ■ Test the motor to see if it will run when connected to the battery.

4. Tell the students to return the materials to the storage area.

How Does a Motor Work? / **97**

LESSON 15

Final Activities Ask the students to describe in their notebooks how they made the motor turn faster, slower, and run in different directions. Encourage them to draw pictures to accompany their explanations.

Extensions
1. Set up a learning center like the one described in **Appendix G**. Invite the students to experiment further, read a book, investigate history, and contribute newspaper articles, drawings, and pictures about the modern (and changing) uses of magnets and motors.

2. After the motors have been reassembled, ask the students to experiment with uses for the motors. Provide the students with "wheels" to attach to the motor shaft (plastic lids work). They also could use rubber bands as belts to turn other "devices."

| LESSON 16 | **Generating Electricity** |

Overview

In this lesson, students explore ways to use the motor as a generator of electricity. They use the generator to operate a motor and light a bulb. They also think about different ways to turn a generator and methods for generating larger amounts of electricity.

Objectives

- Students learn about the source of household electricity.
- Students explore different ways to turn a generator.
- Students learn about another connection between electricity and magnetism.

Background

There is a remarkable symmetry between electricity and magnetism. We saw in Lesson 7, when the compass needle was moved by the electric current in the wire, that electric current has a magnetic field associated with it. The fascinating "other hand" is that, when a wire is moved through a magnetic field, it causes an electric current to flow in the wire. This was first realized in 1830. It made possible the production of large quantities of inexpensive electricity and changed the world dramatically.

The electricity that is generated by electric power companies is produced by the turning of enormous generators. Most commonly, steam is produced by the burning of gas or coal or by a nuclear chain reaction; the steam is then directed through a turbine. The steam makes the turbine rotate and the turbine turns the armature of a generator. A hydroelectric plant uses the weight of falling water to power a turbine and a generator.

In this lesson, when the students turn the shafts of their motors, an electric current is generated in the wires of the electromagnet on the armature. The magnetism "pushes" the electricity, in a sense, along the wire. However, an electric current is created only if the electricity has a circuit to flow through. If the wires are not connected in a circuit, no electricity will flow.

By connecting a light bulb or another motor as part of the circuit, students can see the usefulness of the electricity they generate.

LESSON 16

Materials

For each student
- 1 student notebook
- 1 bulb and bulb socket
- 1 wide rubber band, No. 84
- 1 small, electric motor (Mabuchi RE-260) with wire leads
- 1 plastic drinking straw

Preparation

1. Prepare the materials for distribution.
2. If you have not taught this lesson before, you may want to take some extra time to prepare your presentation of the concept of the generation of electricity in Steps 1 and 2 of the **Procedure** section. This idea is astonishing for many students.

Procedure

1. Begin by asking the class to remind you once again of the basic elements needed for a motor to operate. List their ideas on the board. Try to combine items so that the list contains the essentials. They probably will come up with a list that resembles the following:

Figure 16-1

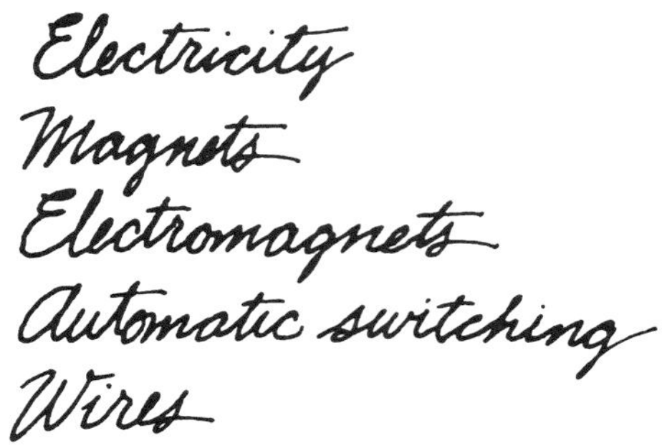

> Electricity
> Magnets
> Electromagnets
> Automatic switching
> Wires

2. Now ask: "If all of these items are combined to make a motor, what do you get from the motor? What does a motor do for you?" Students' responses may range from "run the refrigerator" to "turn" to "work" or "move things." The common thread in all of these is movement. The force provided by the motor causes motion. This connection could be illustrated by adding an arrow to the list of components, as indicated in Figure 16-2.

Figure 16-2

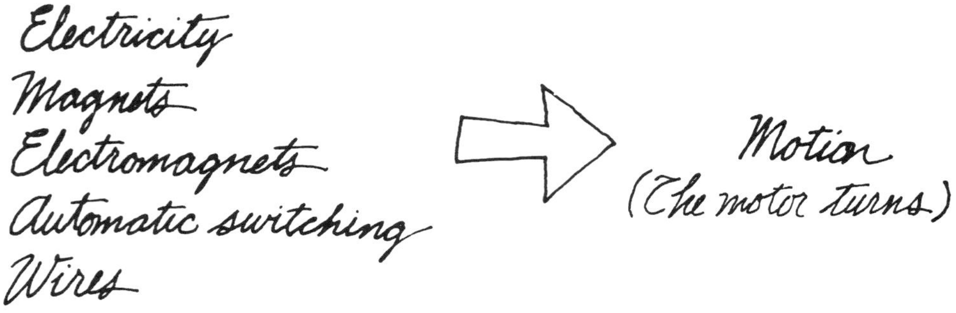

> Electricity
> Magnets
> Electromagnets
> Automatic switching
> Wires   ⇒   Motion
>   (The motor turns)

3. Next, erase or cross out the electricity entry from the list, as demonstrated in Figure 16-3. Replace the word electricity with "turning the motor by hand." Ask the students: "What do you think will happen if we take away

100 / Generating Electricity

the electricity and you turn the motor manually? What will be the result of that?" Illustrate this for the students. Several students may guess that electricity will be created. Ask the students how they would go about finding out for themselves what would happen. After a brief discussion, the students probably will be eager to try turning the motor manually themselves.

Figure 16-3

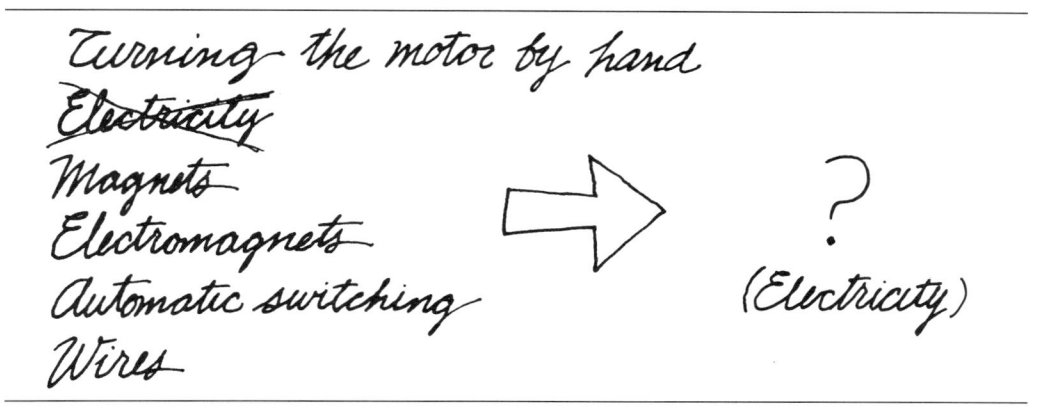

Turning the motor by hand
~~Electricity~~
~~Magnets~~
~~Electromagnets~~
~~Automatic switching~~
~~Wires~~

⇒ ?

(Electricity)

4. Distribute the materials to each student.

5. Ask students to read Steps 3–5 in the Student Activity Book on pgs. 59 and 60 and to try to use the generator to make a bulb light and a motor turn. In Steps 3–5 students are asked to:

   - Hook the motor to the bulb to see if the bulb lights.

   - Try to turn the motor with their fingers while watching the bulb to see if it lights.

   - Wrap a rubber band around the edge of a book, then to run the motor shaft on the rubber band. This should make the motor turn fast and the bulb light. (When the motor is used like this to make electric current flow, it is called a generator.)

6. After most of the students have been successful generating electricity to light the bulb, ask one of the students who has had success to tell the class how it was done. Allow additional time for struggling students to try these suggestions.

7. Tell students that once they have made the bulb light, they should disconnect it from their motor and connect their motor to another student's motor (as shown in Figure 16-3 in the Student Activity Book). Then they should take turns making the motor spin. They can attach the straw to the motor shaft, so that they can see the direction of the spin.

8. Page 61 of the Student Activity Book contains several questions for the students to think about as they work. Remind students to consider these questions:

   - What is the best way you can find to turn the generator?

   - Can you think of a way to use the generator to make the light stay on constantly?

   - What can you do to make the bulb brighter or dimmer?

   - What will make the motor run in a different direction?

   - What else could you use the generator for?

   - Are there different ways to turn the generator?

Generating Electricity / **101**

LESSON 16

9. Ask the students to return the materials to the storage area.

Final Activities

1. Lead a class discussion about the uses of generators and ways to turn them. Some possible questions for getting and keeping the discussion rolling are listed below.

   - How could you generate enough electricity for a whole city?
   - What effects do different methods of turning a large generator have on the environment? (Smoke from coal plants versus radioactivity from nuclear plants, for example.)
   - Are there any ways you could store the electricity for later use?
   - How is a generator like a battery? How are they different?
   - At what time of day do you think the most electricity gets used? Why?

2. Ask the students to write in their notebooks the different ways they can think of to turn a generator. Have them draw a picture of the method they think is best and write an explanation of why they think so.

Extensions

The extensions described below merely touch on the rich possibilities for continuing the processes of discovery and investigation begun in this unit. It is likely that the students will come up with a variety of interesting questions they would like to explore. You may wish to encourage their so-called "self-directed learning" (creating, inventing, experimenting, and researching) by providing additional time for these valuable experiences.

1. Provide the students with additional materials so that they can test their ideas of ways to make the generator turn. A weight on a string will turn the generator fast enough to light a bulb. Blowing or pouring water on a propeller will not, however, provide quite enough rotational speed to light a bulb unless high pressure air or water are used. Let the students discover this for themselves and attempt to find other ways to make the bulb light.

   If you have access to a set of plastic gears (found in motorized toys and some plastic construction toy kits), you may want to let the students experiment with changes in speed accomplished by different gear ratios.

2. Call your local electric utility and arrange to have your class tour an electric generating station. Most power companies maintain an "education office" that provides public relations materials to schools and the general public and that facilitates arrangements for tours.

3. Ask your class to write a story about how their life would be different if electricity were not available.

4. Have the students interview a person who remembers what life was like a long time ago, before electricity was widely available.

102 / Generating Electricity

APPENDIX A

Post-Unit Assessments

Overview

Several possible assessments are included in this appendix. Each provides a different angle on what students have learned.

- **Assessment 1** challenges students to extend what they learned in Lesson 7 and Lesson 12.

- **Assessment 2** is a follow-up to the spinning motor activity that students did in Lesson 13.

- **Assessment 3** asks students to devise a way to solve a measurement problem.

- **Assessment 4** is a post-assessment, identical to the pre-assessment of Lesson 1.

Objectives

- Students review and synthesize what they have learned.
- The teacher assesses the progress of each student.

Background

In Lesson 1, the students began to keep a record in their notebooks of their thoughts and experiences. Now this record will help the students to remember what they have done and allow them to link together the progressive experiences they have had.

You may wish to have the students use their notebooks as they complete the post-unit assessments.

ASSESSMENT 1 An Extension to Lessons 7 and 12

Materials

For each student
- 1 student notebook
- 1 battery and battery holder
- 1 magnetic compass
- 1 piece of #22 coated hook-up wire, 80 cm (32 inches) long
- 2 pieces of #22 coated hook-up wire, 20 cm (8 inches) long
- 1 switch

APPENDIX A

Procedure

1. Challenge students to build and demonstrate an "electric current detector." Using the magnetic compass, they should be able to tell when current is flowing through a wire and when it is not. They may also be able to tell when the current changes direction merely by observing the behavior of the compass when it is placed near the wire. (Students had experience with this phenomenon in Lesson 7.)

2. Ask students to sketch and write explanations of the technique(s) they use.

3. Now ask students to make the magnetic compass needle spin like a motor. Challenge them to accomplish this without moving anything but the switch. (They experienced a similar situation in Lesson 12). Ask students to sketch and explain the technique they use. They should be able to get the needle spinning, but it will be almost impossible for them to keep it moving consistently because it moves so fast.

ASSESSMENT 2   A Follow-up to Lesson 13

Materials

For the class
- 1 demonstration spinning coil motor from Lesson 13
- 2 magnets
- 4 bulbs and bulb sockets
- 2 batteries and battery holders
- 2 alligator leads
- 8 pieces of #22 coated hook-up wire, 20 cm (8 inches) long

Preparation

1. Set up a demonstration of the motor used in Lesson 13. Include four parallel bulbs in series with the motor as shown in Figure A-1.

2. Test to ensure that everything is working properly. The motor should spin continuously and the lights should blink on and off, indicating that the electric current is being interrupted. You may want to use two magnets to help the motor spin more readily.

   Note: At least four bulbs are needed to allow enough electricity to flow through the circuit to make the coil spin.

Procedure

1. Place the spinning coil motor high enough in a central location for all to see. Hook up the motor and ask students to watch closely.

2. Ask students to think of at least two different ways to explain what they see happening. Ask them to write down their ideas, including why they think the lights flash and why they think the coil spins. Their responses may provide insight into how they think about electricity, coils of wire, and magnetism.

Figure A-1

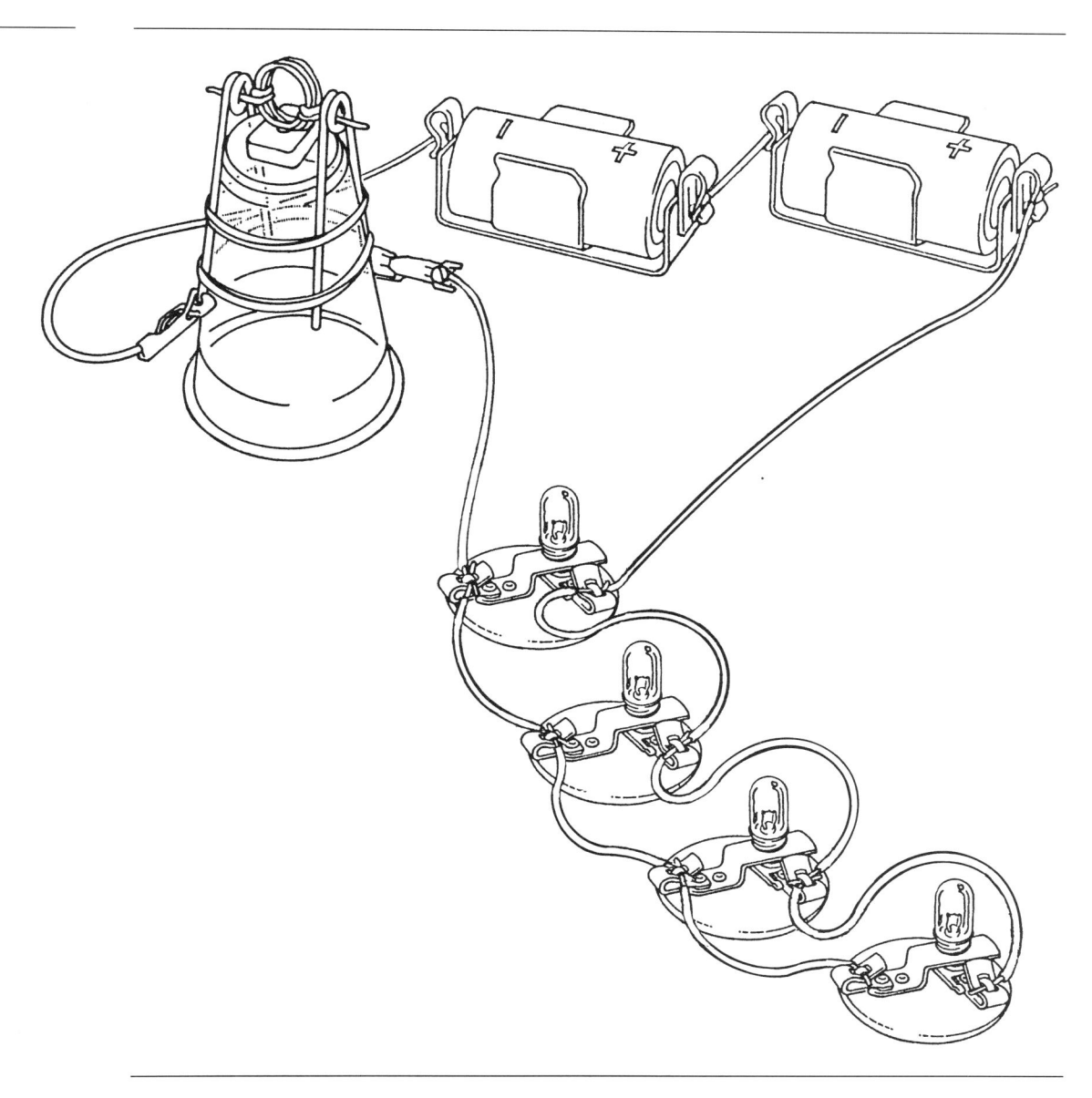

ASSESSMENT 3	**Solving a Measurement Problem**
Materials	*For the class*
	Optional: Provide materials that would help students answer the questions listed in the procedure section below.
Procedure	1. Present students with a measurement problem. Suggestions include:
	■ What new ways can you think of to measure the strength of a magnet?
	■ How could you measure the strength of a horseshoe magnet?
	■ What ways can you think of to count the number of times a motor spins in one minute?

Post-Unit Assessments / **105**

APPENDIX A

- How could you count the number of times a motor spins in one minute if the motor was spinning too fast to see?

You may wish to have the students try to actually test the ideas they have about these questions.

2. Ask the students to construct and/or interpret a graph containing data with which they are familiar. For example:

   - Measure the strength of several different sizes of magnets and report the results on a graph.

   - Present a graph of the magnetic strength of several different combinations of magnets and ask the students to predict, then test, the magnetic strength of each.

ASSESSMENT 4 Post-Assessment to Match Pre-Assessment (Lesson 1)

Materials

For each student
1 copy of **Activity Sheet 6, Post-Unit Assessment**

Procedure

1. Ask students to complete **Activity Sheet 6**.

2. After they have completed the Activity Sheet, allow them to compare the ideas they now have with the ideas they had at the beginning of the unit.

Post-Unit Assessment

Activity Sheet 6

NAME: _____

DATE: _____

What Do You Now Know about Magnets and Motors?

1. Think of all the places that electricity comes from. Make a list of at least five of those places below.

2. Draw a picture of what you think an electric motor looks like. Put labels on your drawing so that other people can see which parts are which.

3. What questions do you still have about magnets, electricity, and motors that you would like to investigate further?

APPENDIX B

Teacher's Record Chart of Student Progress

The Teacher's Record Chart of Student Progress is a convenient way to keep a running record of the progress of individual students. It will help you to track the work students produce as well as the skills they develop and exhibit.

Teacher's Record Chart of Student Progress for *Magnets and Motors*

		Student
Products	Lesson 1: Pre-unit assessment (Activity Sheet 1)	
	Lesson 2: Drawings of "what magnets can do"	
	Lesson 3: Tables of predictions and findings about different materials	
	Lesson 4: Data recorded while measuring magnetic strength	
	Lesson 4: Graph of magnetic strength (Activity Sheet 2)	
	Lesson 5: A compass	
	Lesson 6: Notebook entry about the effect of a magnet on the compass	
	Lesson 7: An electric circuit	
	Lesson 7: Drawing of electric circuit and explanation of compass movement	
	Lesson 8: Two types of electromagnets	
	Lesson 9: Notebook entry about experimental plan	
	Lesson 10: Record of data collected from experiment	
	Lesson 10: Notebook entry about the experimental procedure	
	Lesson 11: Graph of experimental results (Activity Sheet 5)	
	Lesson 11: Notebook entry describing what was learned	
	Lesson 12: Notebook entry explaining how to control the spin of the motor	
	Lesson 13: A spinning coil motor	
	Lesson 14: Written plan of how to reassemble the motor	
	Lesson 15: Drawings and notebook entry about ways to control the motor	
	Lesson 16: Drawing and notebook entry about ways to turn a generator	
Specific Skills	Can demonstrate several characteristics of magnets	
	Can conduct test to distinguish magnetic and non-magnetic materials	
	Can construct a graph to communicate the results of an experiment	
	Can construct and balance a compass	
	Can successfully use a compass to determine geographic direction	
	Can build an electric circuit and investigate the effect of electricity on a magnetic compass	
	Can construct an electromagnet and use it to lift objects	
	Can work cooperatively to plan an experiment	
	Can work cooperatively to conduct a controlled experiment	
	Can use an electromagnetic coil to cause a compass to spin	
	Can construct a spinning coil motor	
	Can successfully disassemble, test, and reassemble an electric motor	
	Can demonstrate the use of a motor to generate electricity	
General Skills	Follows directions	
	Records observations with drawings, words, measurements	
	Works cooperatively	
	Contributes to discussions	

APPENDIX C

Making a Mystery Box

The mystery box is a technique used in Lesson 3 to help students to investigate the properties of magnets and magnetic materials. This appendix provides instructions for making the box. These instructions could be used by students as well as by teachers or adult assistants.

Materials

For every box (one for every two students)
- 1 small, thin, cardboard box (an empty cereal box will work)
- 1 (or more) magnets or magnetic objects to hide in the box, such as:
    - large steel washers
    - magnets
    - nails
    - scissors
    - steel bolts
- 1 copy of **Activity Sheet 7, What Is Inside the Mystery Box?**
- 1 roll of cellophane tape

Procedure

1. Tape a magnet or steel object securely anywhere on the inside of the cardboard box, as shown below.

Figure C-1

Inside a mystery box

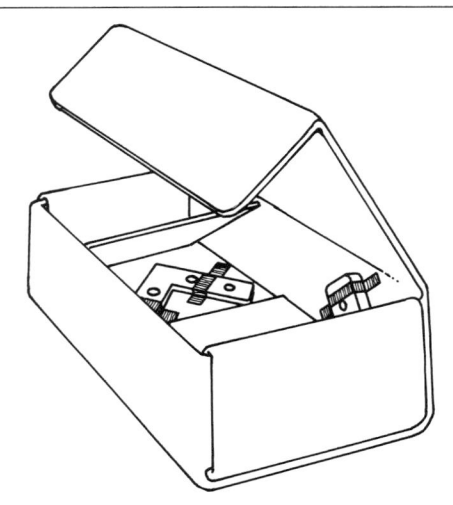

Making a Mystery Box / 111

APPENDIX C

2. Test with a magnet on the outside of the cardboard box to see if you can detect the presence of an object inside. If you cannot locate an object, try using a stronger magnet or a thinner cardboard box. Check to be sure that the object is attracted to a magnet.

3. Seal the box and wrap **Activity Sheet 7** firmly around it. Tape or glue the activity sheet in place.

4. Mark the box with a number so that students who use it can identify it and compare it with other mystery boxes.

Figure C-2 shows a mystery box ready for use.

Figure C-2

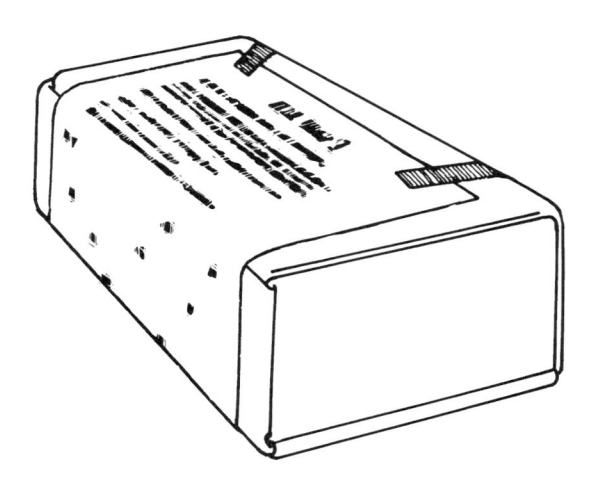

A mystery box ready for use

112 / Making a Mystery Box

APPENDIX C

What Is Inside the Mystery Box? Activity Sheet 7

Box Number _____

Write down your responses to these questions in your notebook. Write down the box number so that you will be able to compare your answers with those of other students.

1. Try to find the object without opening the box. Where is the object? (What letter is it closest to?)
2. Is the object magnetic? How do you know?
3. Is the object a magnet? What makes you think so?
4. What shape do you think the object is?
5. Are there other objects hidden? How many? What evidence did you find?

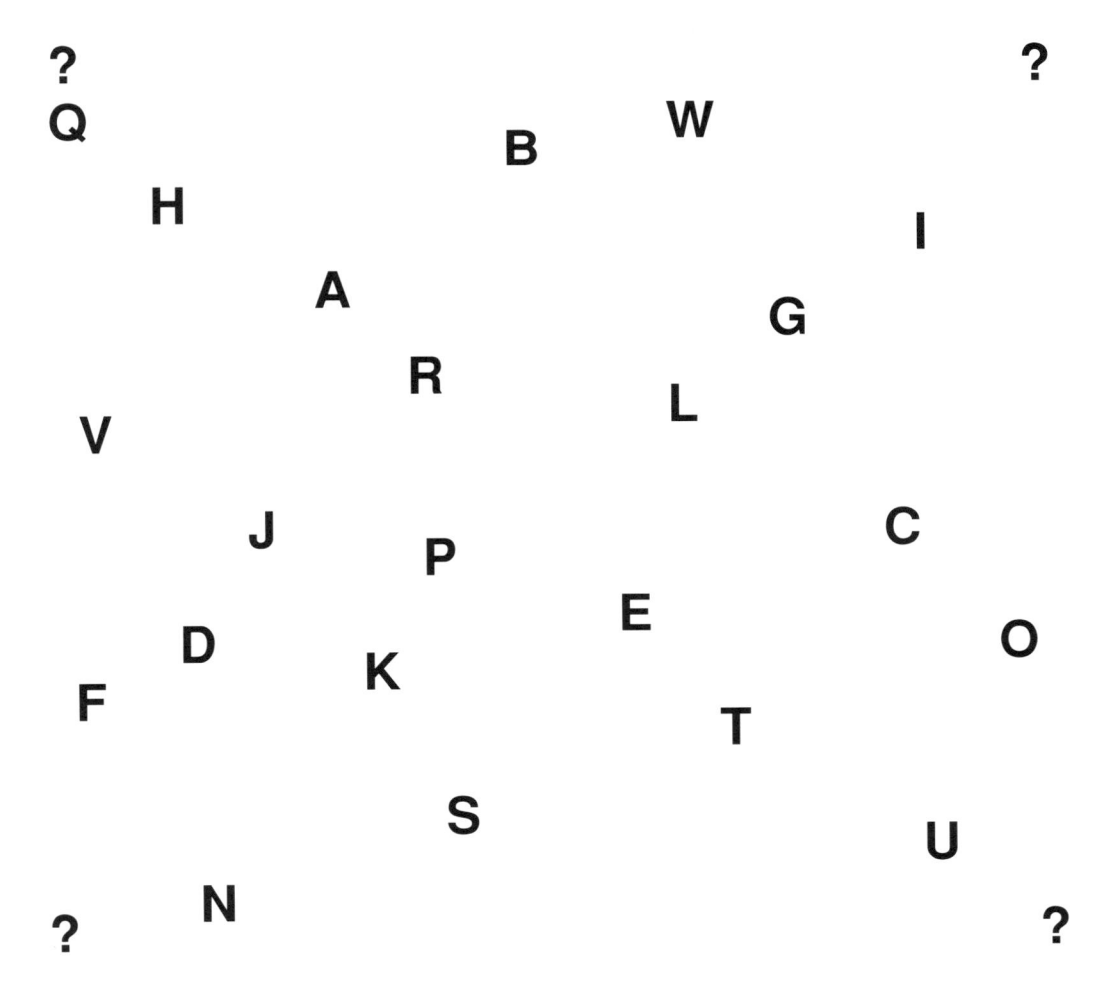

```
  ?                                    ?
  Q                       W
              H       B
                  A           I
                          G
                  R
                      L
  V
          J
              P         C
                    E
          D               O
      F       K
                      T
              S
                          U
  ?       N                     ?
```

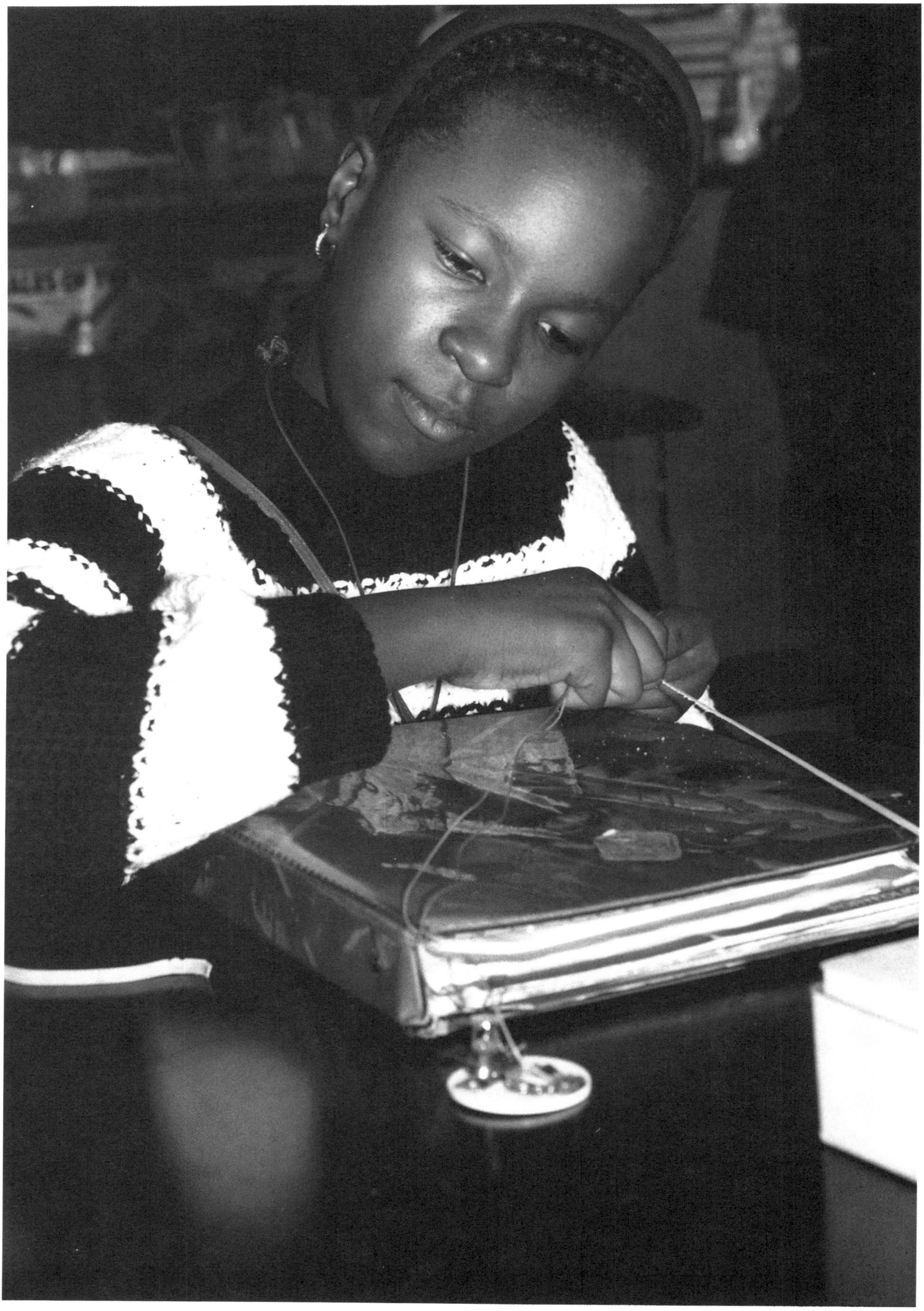

APPENDIX D

Background: Electric Circuits

The following paragraphs describe the basic components of an electric circuit and explain the differences between series and parallel circuits.

Electric Current in a Circuit

Electricity flows along a path called a circuit. To create a circuit, you need a battery, a conductor, and whatever else you wish to include in the circuit, such as a switch, a motor, and/or a bulb. The electricity must be able to move from one end of the energy source (in this case, a battery) to the other to be able to create a complete circuit.

How does electricity flow along a circuit? Like many things in nature, it is invisible, but we can see and measure the results of the flow. The battery, or energy source, gives electricity its "push" through a circuit. This push, or voltage, can be thought of as electrical pressure and is analogous to water pressure. Electrical pressure is measured in volts.

The flow of electricity through a circuit is something like the flow of water through a hose. The flow of electrical current is measured in amperes.

By convention, scientists and engineers speak of the electric current in a circuit as though it flowed from the positive end of the battery through the wires and back to the negative end of the battery. This convention arose because of Benjamin Franklin's theory that electric current was carried by positive charges. As it turned out, Franklin's theory has been proven incorrect. Today, we know that it is the movement of electrons, which have a negative charge, that constitutes an electric current. But speaking of electric current as if it were the movement of positive particles has been preserved as a convention.

When explaining the flow of electricity to students, it probably is less confusing to focus on this conventional current (flow from positive to negative). But some students may know about electron flow already and bring it up during the discussion.

Basic Parts of a Circuit

In this unit, students will be working with a D-cell battery. The D-cell, like all batteries, has two ends, one marked + (positive) and one marked – (negative). The positive end has a small, raised button on it; the negative end is flat.

APPENDIX D

The wire that students will be working with is commonly called "hook-up wire" or bell wire because it is used to connect doorbell circuits. It usually has a conducting center made of copper or aluminum. The conducting wire allows electricity to flow from one end of the battery to the other. In some other kinds of wires, such as lamp cords, there are two strands of wire inside. The two strands are necessary so that a complete circuit can be created, beginning at the power plant, going through a network of wires to an appliance or light, and back to the power plant. To prevent shocks, the wire is covered by a plastic sheath that acts as an insulator.

The bulbs used in this unit are very similar to the household bulbs in fixtures and lamps, except that they are much smaller. A typical bulb is shown in Figure D-1; its parts have been labeled. The filament is the part of the bulb that gets hot, glows, and produces light.

Figure D-1

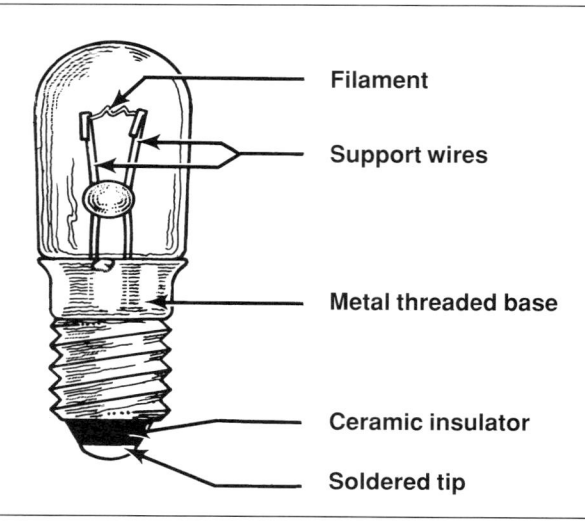

- Filament
- Support wires
- Metal threaded base
- Ceramic insulator
- Soldered tip

Sometimes, students connect a wire directly from one end of the battery to the other without having a bulb or motor in the circuit. When this happens, a short circuit occurs.

Short circuits in a house or an automobile can produce dramatic sparks and enough heat to melt metals and start a fire. Short circuits with D-cell batteries are not dangerous, but they do drain the electrical energy from the batteries. If left connected for several minutes, both the wire and the battery will get warm.

In Lesson 7, students are asked to build a kind of short circuit. Please be sure that they always include a switch in the circuits so that the electricity can be turned off easily. If the electricity is left flowing through a short circuit for longer than a few seconds at a time, the battery will not last as long as it normally would.

Kinds of Circuits

There are two kinds of circuits: series circuits and parallel circuits. In a series circuit, electricity has only one path to travel, from its starting point on the circuit through the wires, batteries, and bulbs and back to its starting point. Figure D-2 shows two batteries arranged in series.

When batteries are arranged in series, the voltage across the bulb is increased, causing the bulb to glow more brightly than it did when only one

APPENDIX D

battery was used. But the batteries will drain more quickly than they would in a parallel arrangement.

In a parallel circuit, electricity can travel along more than one path around the circuit. Figure D-3 shows two batteries arranged in parallel. Note that there are two paths through which electricity can flow around the circuit.

When batteries are arranged in parallel, the voltage across the bulb remains the same. The brightness of the bulb will be the same as it was with one battery. But the bulb will burn longer in this circuit than it will when the batteries are arranged in series.

Figure D-2

Two batteries in series, one bulb

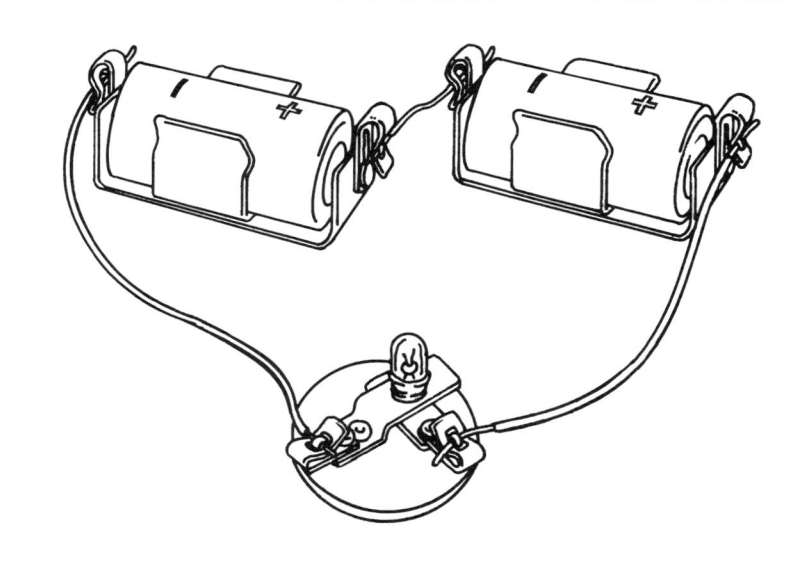

Figure D-3

Two batteries in parallel, one bulb

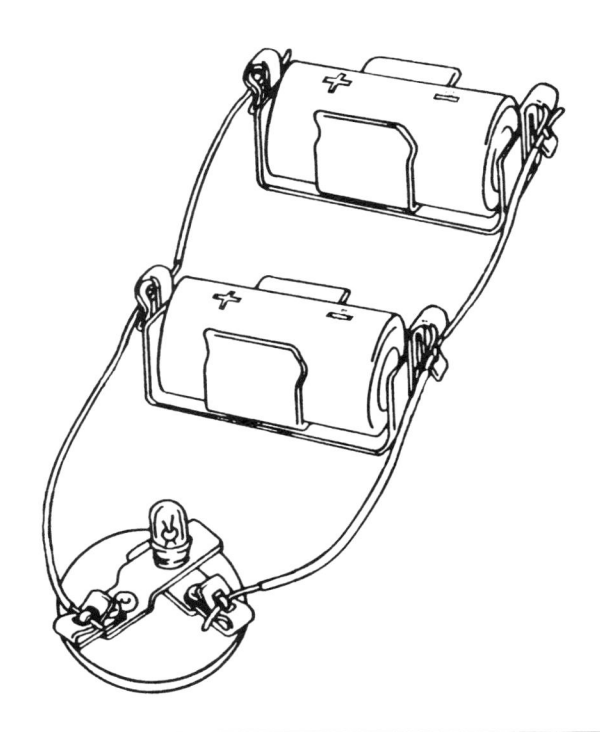

Background: Electric Circuits / **117**

APPENDIX D

Bulbs can be wired in series or in parallel, as well. When two identical bulbs are wired in series with one battery, they burn with uniform brightness, but more dimly than when only one bulb is connected to the battery. Because they are part of the same circuit, if one is removed from its socket, the other will also go out. Figure D-4 shows bulbs arranged in series.

Figure D-4

Two bulbs in series, one battery

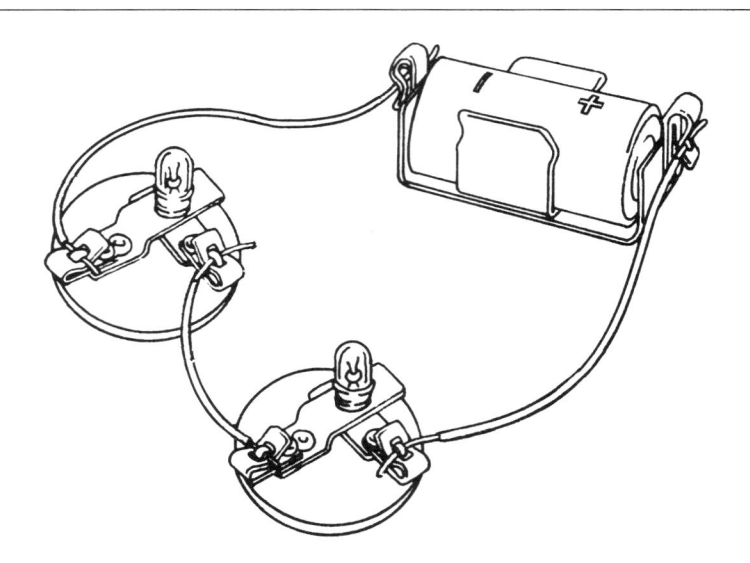

When two bulbs are wired in parallel with one battery, each bulb burns as brightly as it did in a one-bulb/one-battery configuration. In addition, when bulbs are arranged in a parallel circuit, unscrewing one bulb does not make the other bulb go out. This is because the electricity travels in independent paths through each bulb. Figure D-5 shows bulbs arranged in parallel.

Figure D-5

Two bulbs in parallel, one battery

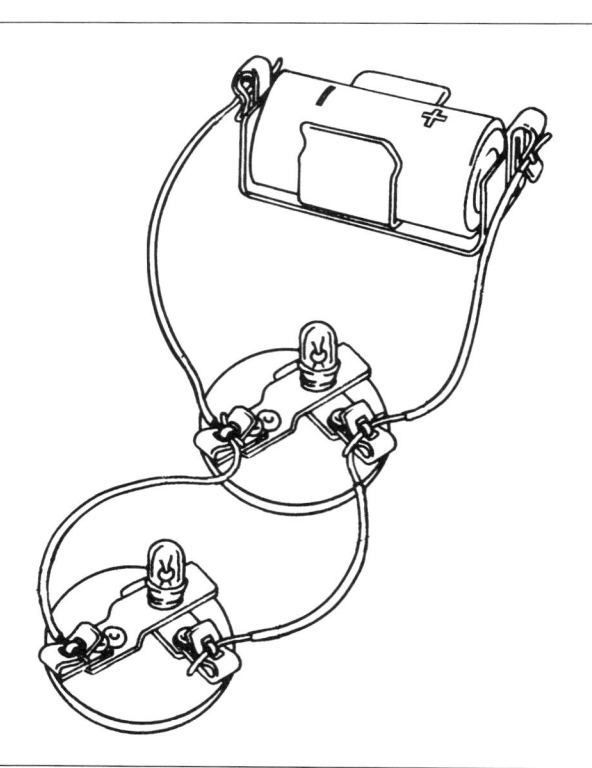

118 / Background: Electric Circuits

APPENDIX E

Using a Wire Stripper

The hook-up wire referred to in materials lists throughout this unit is covered by a plastic, insulating sheath. To expose the metal conductor, strip away the cover with a wire-stripping tool. (A small knife or wire cutter also can be used.) Use the same tool to cut the wire.

Figure E-1 shows one stripping tool and how to use it. The upper jaws are used to strip the wire. Place the wire between the jaws and squeeze the handle. The jaws grip the wire. The blades should cut through the insulation only, and then strip it from the wire. To use the wire cutter on the tool, place the wire between the lower cutters, and squeeze the handle. Another type of wire-stripping and cutting tool is shown in Figure E-2.

APPENDIX E

Figure E-1

Wire-stripping and wire-cutting tool

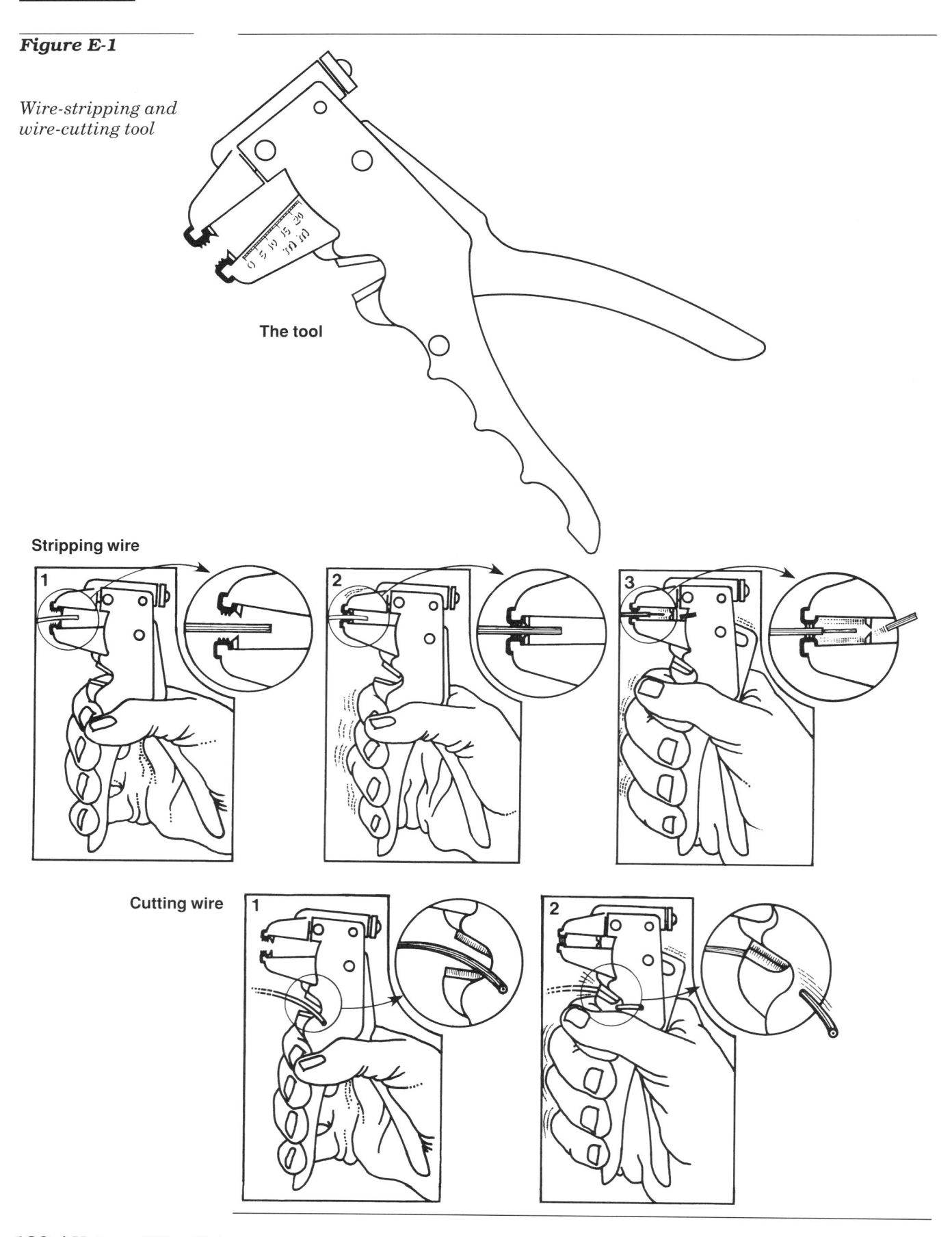

The tool

Stripping wire

Cutting wire

120 / Using a Wire Stripper

APPENDIX E

Figure E-2

Another wire-stripping and wire-cutting tool

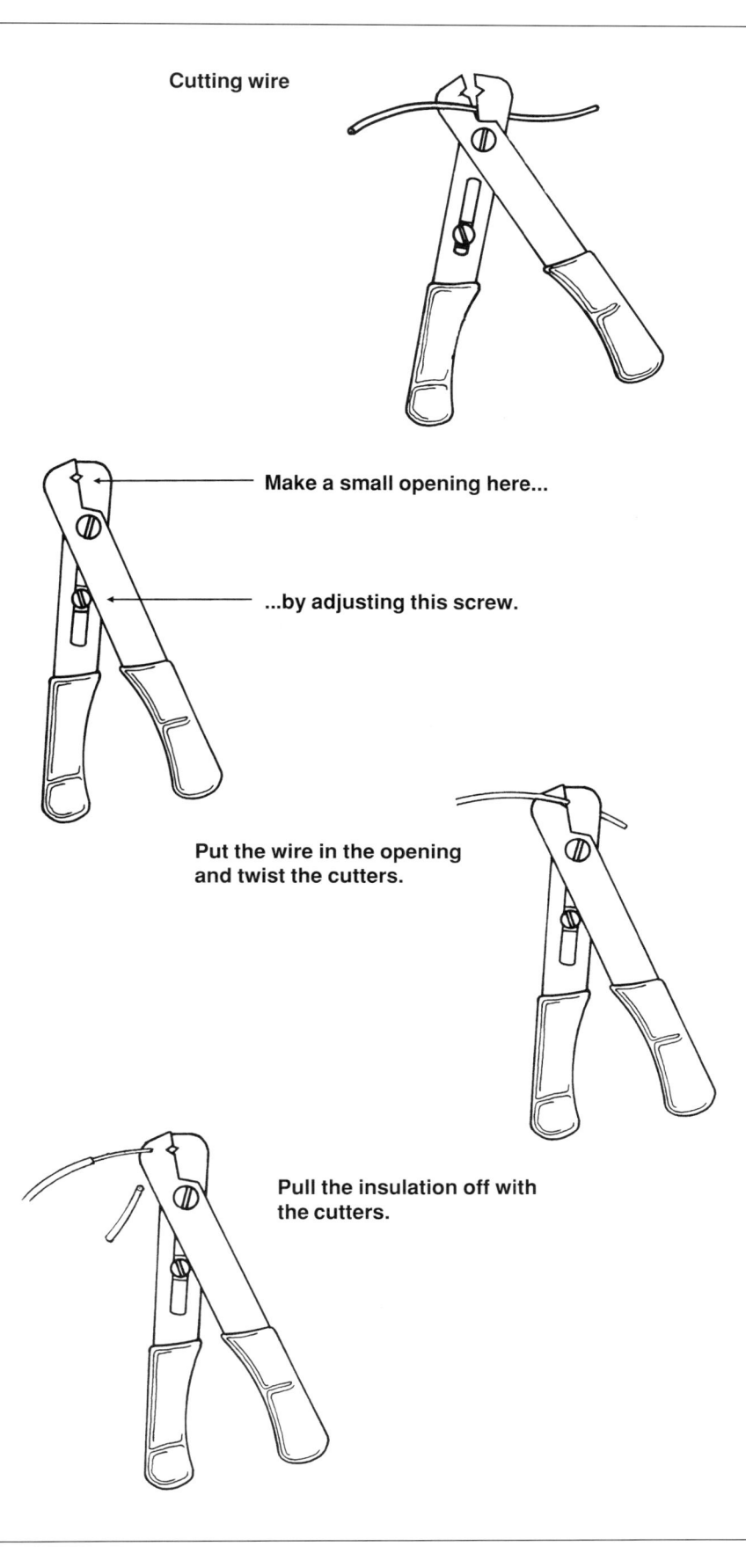

Cutting wire

Make a small opening here...

...by adjusting this screw.

Put the wire in the opening and twist the cutters.

Pull the insulation off with the cutters.

Using a Wire Stripper / **121**

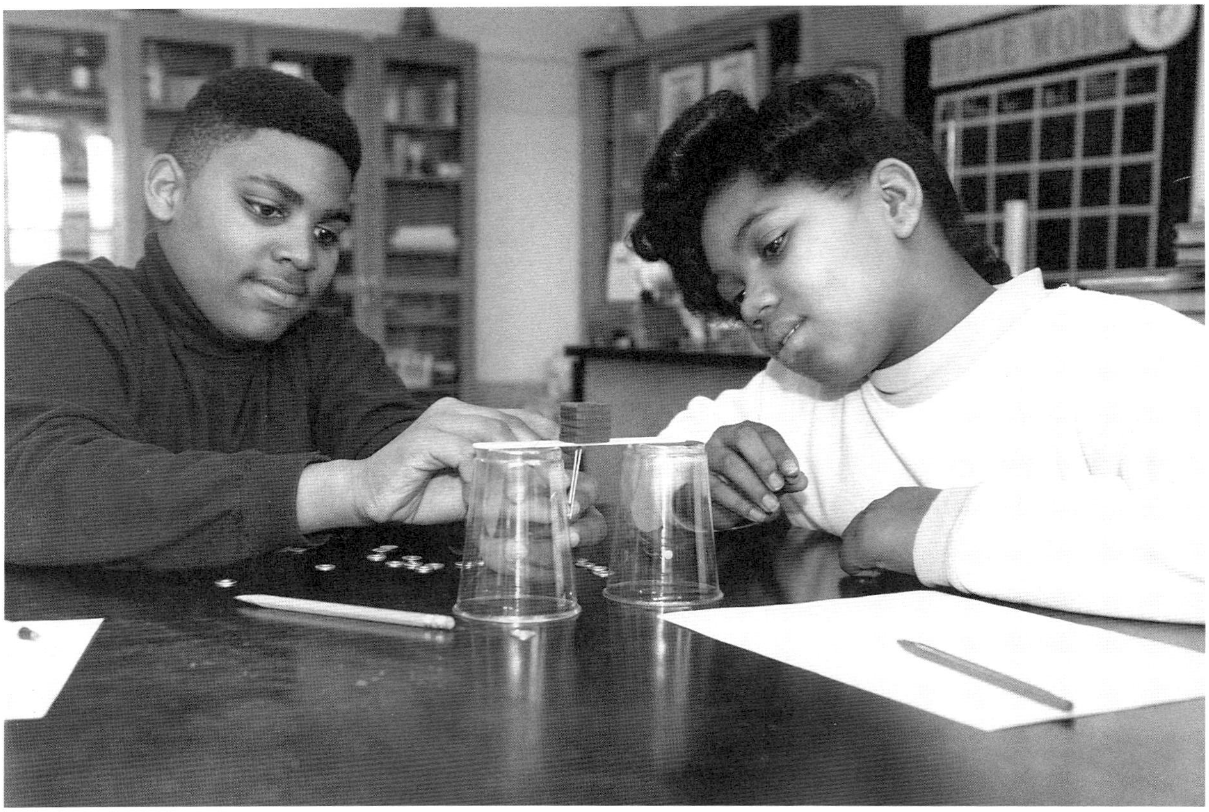

APPENDIX F

Making and Repairing Alligator Leads

To make an alligator lead or to repair one that is no longer functioning, loosen the screw on the alligator clip two or three turns with a screwdriver.

Next, strip the insulation off the end of a piece of wire (see **Appendix E**) and bend the bare end around the screwdriver, as shown in Figure F-1.

Figure F-1

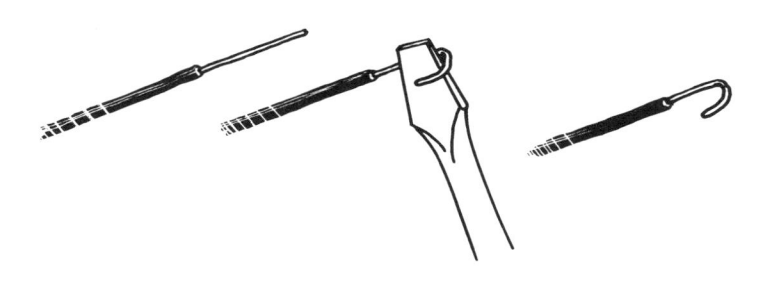

Hook the end of the wire around the screw on the alligator clip, then tighten the screw by turning it clockwise with the screwdriver as shown in Figure F-2. The head of the screw will hold the wire to the alligator clip.

Note: Do not bend the alligator clip around the wire or it will be difficult to repair if the wire breaks.

Figure F-2

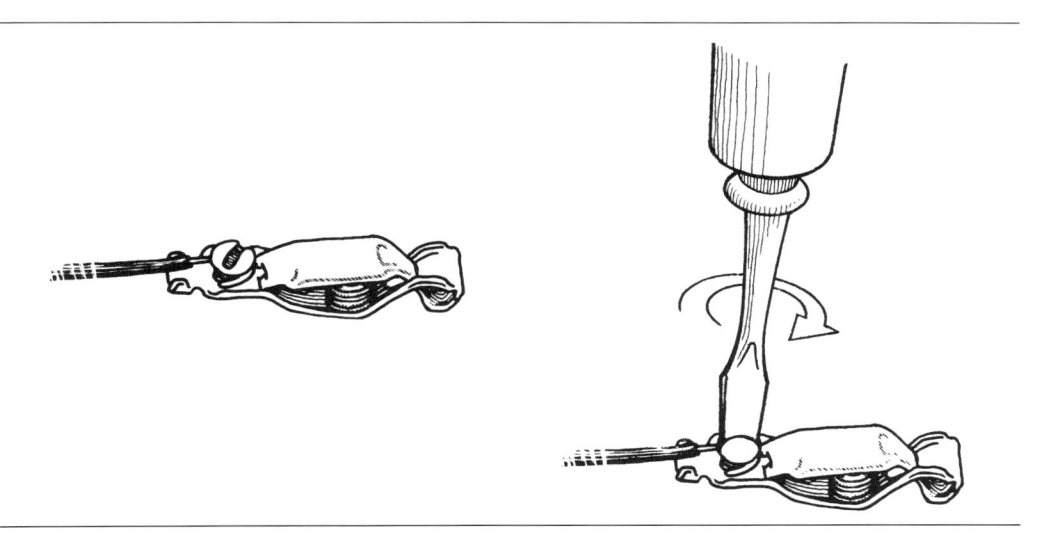

APPENDIX G

Setting up a Learning Center for *Magnets and Motors*

Here are a few suggestions for setting up a learning center for this unit, an example of which is shown in the drawing below.

Figure G-1

The components of such a learning center can be:

- A timeline (see pg. 129) and map, which offer a historic and geographic overview.
- Trade books on electricity and magnetism for students to check out and read.

APPENDIX G

- Articles about current events and scientific breakthroughs involving electricity and magnetism.
- Tidbits of information about electromagnets and permanent magnets (see pgs. 126 through 128).
- Students' ideas about magnets and motors.
- Suggestions of projects and materials for students to use to explore and experiment further with magnets, electricity, and motors.

The learning center can be built from a large cardboard box or a bulletin board. Ask your students for additional ideas about what should be included.

The next several pages are examples of tidbits of information for display in the learning center.

Electromagnets...

are made by electric current moving though a coil of wire wrapped around an iron, steel, or alloy core.

If the coil of wire is not wrapped around a solid core, sometimes it is called a **solenoid**. Usually a **solenoid** is a cylindrical coil.

Figure G2 and G3

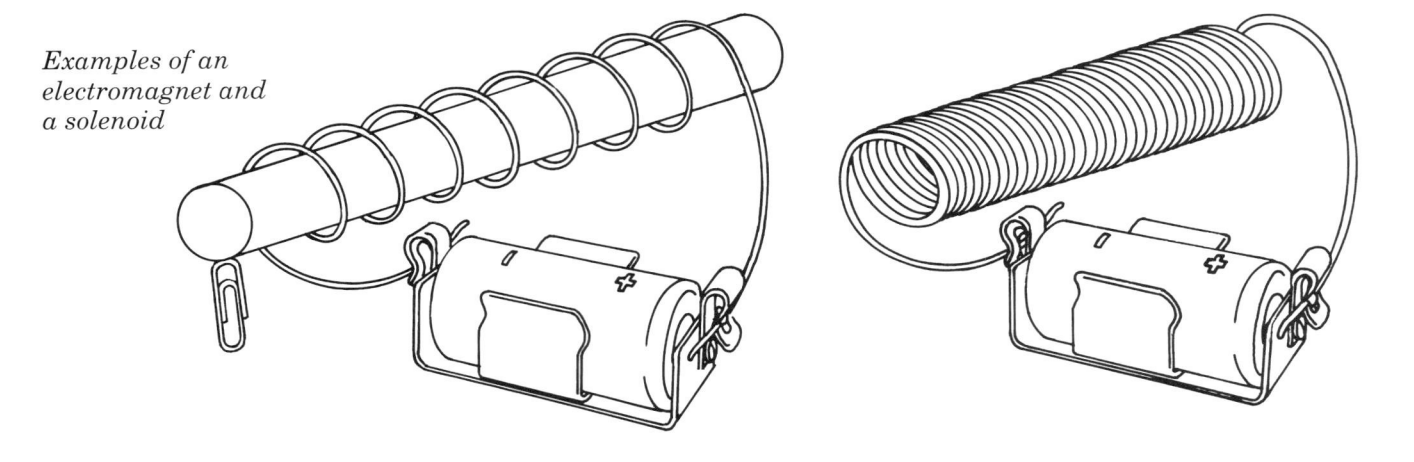

Examples of an electromagnet and a solenoid

Permanent magnets...

are made in several ways:

1. **Magnetite** is a kind of rock that is naturally magnetic. Magnetism is produced by the earth's magnetic field while the rock is still hot. When the rock cools, it stays magnetic.

2. Stroking an object that is made of iron, nickel, or cobalt with a magnet makes the object become magnetic.

3. Companies that produce magnets use powerful electromagnets to make permanent magnets. They heat up pieces of iron, nickel, or cobalt and then use the electromagnets to keep the materials magnetic while they cool off. When the materials cool off, they stay magnetic—much like the magnetite in No. 1.

Figure G-4

Magnetite is naturally magnetic

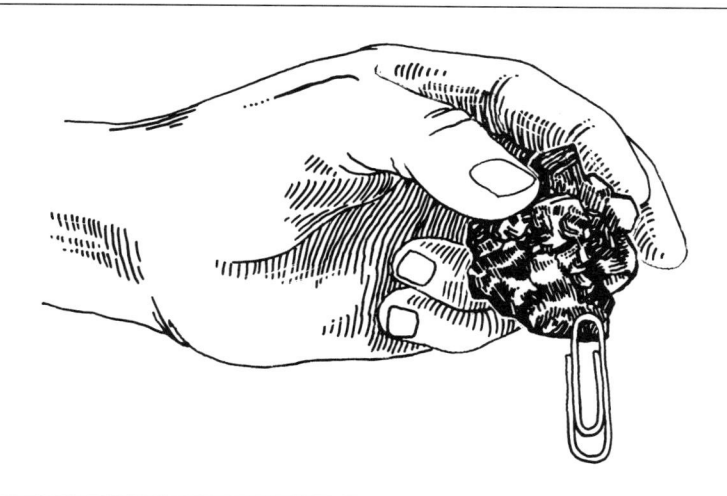

Figure G-5

Stroking a nail to make it become magnetic

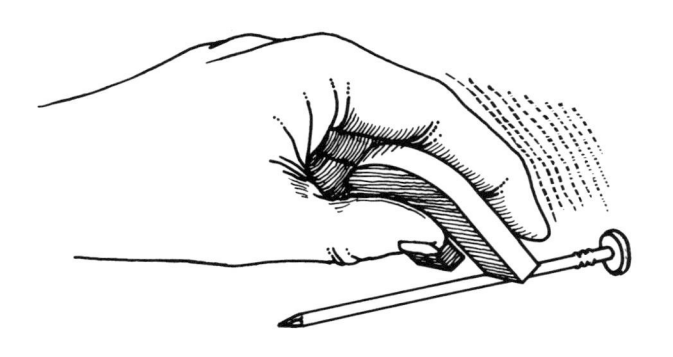

Can you build...

... a Circuit Breaker?

Description: A circuit breaker is an automatic switch that turns electricity off when the current gets too high. This prevents the wires from getting so hot that they will cause a fire.

Hint: Remember that a coil of wire will become magnetic when electricity flows through it. Use the push of the magnetic coil to pull the switch open and stop the flow of the electric current.

Can you build...

... an Electric Doorbell?

Description: A doorbell can be anything that makes a sound when you turn a switch on.

Hint: Remember that a coil of wire will become magnetic when electricity flows through it. Use the push of the magnetic coil to project something into a bell to make it ring.

APPENDIX G

Can you build...

... a Speaker?

Description: A speaker is a device that converts electric signals into sound. Radios and televisions use speakers to produce the sound that the listener hears.

Hint: You can get the electric signals from a portable radio by plugging an old earphone jack into the radio's earphone hole. Try taping a magnet to a coil of wire on the bottom of a plastic cup. The changing electric current makes the magnet vibrate, and the magnet and wire make the cup vibrate—so you hear the vibrations!

Can you build...

... a Color Wheel?

Description: A color wheel will let you draw circles as long as the motor is running.

Hint: Poke the motor shaft through the center of a plastic lid. Tape a piece of paper to the lid and turn the motor on. Try drawing on the paper with a pencil or marker pen.

Magnetism and Electricity:
When Important Discoveries Were Made

600 BC	About this time, Thales of Miletus wrote about magnetite attracting bits of iron.
1000 AD	About this time, the first magnetic compass appeared in China.
1492	During his first voyage to the Americas, Christopher Columbus discovered that the magnetic compass doesn't always point due north.
1600	William Gilbert suggested that the Earth is a huge, spherical magnet.
1749	John Canton developed a method for making an artificial magnet.
1819	Hans Christian Oersted discovered that a compass needle is deflected by electric current.
1821	Michael Faraday built the first electric motor.
1823	William Sturgeon built the first electromagnet.
1830	Joseph Henry discovered the principle of the electric generator.
1930	A tape recorder was developed that uses magnetic plastic tape.

APPENDIX H

Bibliography

Resources for Teachers

Dishon, Dee, and O'Leary, Pat Wilson. *A Guidebook for Cooperative Learning: Techniques for Creating More Effective Schools.* Holmes Beach, Florida: Learning Publications, Inc., 1984. (Cat. No. 97-2684)*

> A practical guide for teachers who are embarking on the implementation of cooperative learning techniques in the classroom.

James, Portia P. *The Real McCoy: African-American Invention and Innovation, 1619-1930.* Washington, D.C.: Smithsonian Institution Press, 1989. (Cat. No. 97-2688)

> *The Real McCoy* tells of the creative spirit of black men and women from the time of the earliest settlements through slavery and emancipation to modern times. The electrical inventions of Granville Woods and Lewis H. Latimer are dealt with in the last pages of the book. While the individual accounts are brief, the photographs and drawings provide a unique historical context for their work.

Johnson, David W., Johnson, Roger T., and Holubec, Edythe Johnson. *Circles of Learning.* Alexandria, VA: Association for Supervision and Curriculum Development, 1984. (Cat. No. 97-2686)

> This excellent book presents the case for cooperative learning in a concise and readable form. It reviews the research, outlines implementation strategies, provides definition to the skills needed by students to work cooperatively, and answers many questions.

Long, Michael E. "Secrets of Animal Navigation." *National Geographic,* Vol. 179, No. 6 (June 1991). pp. 70-99.

> This article on the secrets of animal navigation provides an overview of what scientists currently know and are still investigating on this subject. The striking photographs will draw students in.

MacAuley, David. *The Way Things Work.* Boston: Houghton, Mifflin Co., 1988. (Cat. No. 97-3470)

> The relevant sections of this book deal with electromagnetism, the production of electricity, and electromagnetic devices. The book

features outstanding pictorial descriptions of various phenomena. The text is fanciful at times, which may confuse children if they read it alone.

Math, Irwin. *Wires and Watts: Understanding and Using Electricity.* New York: Charles Scribner's Sons, 1981. (Cat. No. 97-2690)

The book presents the fundamentals of electricity through experiments and projects that produce actual working models. It goes considerably beyond the content of *Magnets and Motors*, introducing the use of simple equations to express Ohm's law, detailing the working of a D-cell battery, and explaining a variety of instruments and devices. This book would be of interest to the most advanced and ambitious students, who could use it with an adult.

Resources for Students

Chapman, Philip. *Electricity (The Young Scientist Book Of).* Tulsa: Educational Development Corp., 1976. (Cat. No. 97-2672)

This book gives some interesting details about how electricity is generated, and a brief account of batteries. It shows a detailed cutaway view of the inside of a D-cell. The book goes well beyond the scope of this unit, and may answer, briefly and concisely, many of the questions that come up.

Cosner, Sharon. *The Light Bulb.* New York: Walker and Company, 1984.

This is a serious book for young readers who are interested in the story of Edison's invention of the light bulb. It provides enough of the details of his struggles and the gritty work involved to give the budding young inventor a realistic view of the tasks that are ahead.

Davidson, Margaret. *The Story of Alexander Graham Bell, Inventor of the Telephone.* New York: Dell Publishing, 1989. (Cat. No. 97-3471)

This biography portrays the human side of an ingenious person. The story is well paced and shows an exciting side of science—discovery. The roles of magnetism and electricity in the development of the telephone are discussed clearly.

Davidson, Margaret. *The Story of Benjamin Franklin, Amazing American.* New York: Bantam Doubleday Dell, 1988. (Cat. No. 97-2676)

This is an excellent, lively biography. It captures Franklin's curiosity and inventiveness while maintaining the very human dimensions of the man. This book is worth reading both for the science and for the history.

Fritz, Jean. *What's the Big Idea, Ben Franklin.* New York: Putnam Publishing Group, 1976. (Cat. No. 97-2678)

This charming biography is excellent for the history and for the down-to-earth qualities it captures of Franklin's life. His systematic problem-solving, which led to many useful inventions, is very well presented. This book is worth reading both for the science and for the literary value.

Gutnik, Martin J. *Michael Faraday.* Chicago: Children's Press, 1986. (Cat. No. 97-3472)

> This is an interesting biography of an important but little-known scientist. The details of his accomplishments are sometimes tedious, but the historical information presented is worthwhile. The black-and-white illustrations add a historical flavor to the story.

Haber, Louis. *Black Pioneers of Science and Invention.* New York: Harcourt Brace Jovanovich, Inc., 1970. (Cat. No. 97-2692)

> This biographic compilation describes the lives and contributions of 14 black American scientists and inventors. Among them, Granville Woods and Lewis Latimer are best known for their work with electricity and magnetism.

Hogan, Paula. *The Compass.* New York: Walker and Company, 1982. (Cat. No. 97-3473)

> This book traces the history of navigation and the technological advances made on magnetic compasses. This book is part of the *Inventions that Changed Our Lives* series.

Whyman, Kathryn. *Electricity and Magnetism (Science Today Series).* New York: Gloucester Press, 1986. (Cat. No. 97-3475)

> This book uses magnificent color photographs and illustrations to explain the workings of magnetism, electricity, and electromagnetic devices. It is an excellent review of the concepts discussed in this unit.

Note: These books may be available from your local public or school libraries, bookstores, or their publishers. They also are available from Carolina Biological Supply Company. To order them from Carolina, use the numbers in parentheses labeled, "Cat. No." Place orders by calling, toll free, 800-334-5551. In North Carolina, call 800-632-1231.

APPENDIX I

Materials Reorder List

97-3402	Teacher's Guide
97-3403	Student Activity Book
97-3420	flexible magnets, 25 x 20 x 5 mm (1" x ¾" x 3⁄16") with a 5-mm (3⁄16-inch) hole
97-3421	plastic cups
97-3422	plastic lids
97-3423	spool of light string
97-3424	pencils, No. 2
97-3425	wooden sticks, 15 cm x 4 mm (6" x ⅛")
97-3426	tongue depressors
97-3427	boxes, cardboard, 10 x 5 x 20 cm (4" x 2" x 8")
97-3428	packages of assorted objects, each containing:
	steel washer
	steel nail
	aluminum screen
	aluminum foil
	brass fastener
	brass washer
	rubber band
	paper clip
	copper wire
	pipe cleaner
	recording tape
	golf tee
	aluminum wire
	twist-tie
97-3429	jumbo paper clips
97-3430	No. 1 paper clips
97-3431	steel washers, USS standard No. 10
97-3432	magnetic compasses
97-3433	straight pins, 2.5 cm (1")
97-3434	plastic drinking straws
97-3435	single-color stickers

APPENDIX I

97-3436	D-cell alkaline batteries
97-3437	battery holders
97-3438	bulbs
97-3439	bulb sockets
97-3440	wire-stripper tool
97-3441	aluminum nails, 12D
97-3442	nails, common 12D
97-3443	brass bolts, 8 cm x 6 mm (3" x ¼")
97-3444	steel bolts, 12 mm x 8 mm (½" x ⅜")
97-3445	steel bolts, 8 cm x 6 mm (3" x ¼")
97-3446	steel bolts, 3.5 cm x 6 mm (1¼" x ¼")
97-3447	steel bolts, 8 cm x 4 mm (3" x ⅛")
97-3448	steel bolts, 8 cm x 8 mm (3" x ⅜")
97-3449	steel bolts, 8 cm x 12 mm (3" x ½")
97-3450	steel bolts, 10 cm x 6 mm (4" x ¼")
97-3451	steel bolts, 13 cm x 6 mm (5" x ¼")
97-3452	graph grid transparencies
97-3453	transparency marker pens
97-3454	envelopes
97-3455	construction paper, assorted colors
97-3456	alligator clips
97-3457	roll (23 m, 50 feet) #20 bare wire
97-3458	rolls (7.5 m, 25 feet) #20 coated hook-up wire, 1 roll each of two different colors
97-3459	roll (30 m, 100 feet) #22 coated hook-up wire (90 m)
97-3460	roll (23 m, 75 feet) #28 enameled wire
97-3461	sandpaper squares, 5-cm (2-inches) square
97-3462	screwdriver
97-3463	small, electric motor, Mabuchi RE-260, with wire leads (30 cm, 12 inches)
97-3464	roll of PVC electrical tape
97-3465	rubber bands, No. 16
97-3466	rubber bands, No. 84

Order materials from:

Carolina Biological Supply Company
2700 York Road
Burlington, NC 27215

Call toll free 800-334-5551.
In North Carolina, call 800-632-1231.

National Science Resources Center Advisory Board

Chairman

David Sheetz, Consultant, The Dow Chemical Company, Midland, Michigan

Members

Bruce M. Alberts, Professor of Biochemistry and Biophysics, University of California at San Francisco, San Francisco, California

Albert V. Baez, President, Vivamos Mejor/USA, Greenbrae, California

Marjory Baruch, Educational Consultant, Fayetteville, New York

Ann Bay, Director, Office of Elementary and Secondary Education, Smithsonian Institution, Washington, D.C.

DeAnna Banks Beane, Project Director, YouthALIVE, Association of Science-Technology Centers, Washington, D.C.

F. Peter Boer, Executive Vice President, W.R. Grace & Company, New York, New York

Martha A. Darling, Manager, Government Policy, The Boeing Company, Seattle, Washington

Hubert M. Dyasi, Director, Workshop Center for Open Education, City College of New York, New York, New York

James D. Ebert, Vice President, National Academy of Sciences, Washington, D.C.; Director, Chesapeake Bay Institute, Johns Hopkins University, Baltimore, Maryland

Douglas E. Evelyn, Deputy Director, National Museum of American History, Smithsonian Institution, Washington, D.C.

Robert M. Fitch, Senior Vice President, retired, S.C. Johnson Wax, Racine, Wisconsin

Samuel H. Fuller, Vice President, Corporate Research, Digital Equipment Corporation, Maynard, Massachusetts

Charles N. Hardy, Assistant Superintendent, Instruction and Curriculum, Highline School District, Seattle, Washington

Martin O. Harwit, Director, National Air and Space Museum, Smithsonian Institution, Washington, D.C.

Robert M. Hazen, Staff Scientist, Carnegie Institution of Washington, D.C.

Robert S. Hoffmann, Assistant Secretary for the Sciences, Smithsonian Institution, Washington, D.C.

Ann P. Kahn, Director, Organizational Liaison, Mathematical Sciences Education Board, National Research Council, Washington, D.C.

Manert Kennedy, Executive Director, Colorado Alliance for Science, University of Colorado, Boulder, Colorado

Sarah A. Lindsey, Science Coordinator, Midland Public Schools, Midland, Michigan

Thomas E. Lovejoy, Assistant Secretary for External Affairs, Smithsonian Institution, Washington, D.C.

William J. McCune, Jr., Chairman, Polaroid Corporation, Cambridge, Massachusetts

Phylis R. Marcuccio, Assistant Executive Director for Publications, National Science Teachers Association, Washington, D.C.

Lynn Margulis, Professor of Biology, University of Massachusetts, Amherst, Massachusetts

Philip Morrison, Professor of Physics, Emeritus, Massachusetts Institute of Technology, Cambridge, Massachusetts

Phylis Morrison, Educational Consultant, Cambridge, Massachusetts

Philip Needleman, Corporate Vice-President, Research and Development, and Chief Scientist, Monsanto Company, St. Louis, Missouri

Jerome Pine, Professor of Physics, California Institute of Technology, Pasadena, California

Wayne E. Ransom, Executive Director of Educational Programs, Franklin Institute, Philadelphia, Pennsylvania

Peter H. Raven, Director, Missouri Botanical Garden, St. Louis, Missouri

Lynne Y. Strieb, Elementary School Teacher, Greenfield School, Philadelphia Public Schools, Philadelphia, Pennsylvania

Melvin R. Webb, Dean, School of Education, Clark Atlanta University, Atlanta, Georgia

Paul H. Williams, Director, Center for Biology Education; and Professor, Department of Plant Pathology, University of Wisconsin, Madison, Wisconsin

Karen L. Worth, Faculty, Wheelock College, Boston, Massachusetts; Senior Associate, Urban Elementary Science Project, Education Development Center, Newton, Massachusetts

Ex Officio Members

James C. Early, Assistant Secretary for Education and Public Service, Smithsonian Institution, Washington, D.C.

Philip M. Smith, Executive Officer, National Academy of Sciences, Washington, D.C.